WES JACKSON

Altars of

Unhewn Stone

SCIENCE AND THE EARTH

NORTH POINT PRESS
Farrar, Straus and Giroux
New York

Copyright © 1987 by Wes Jackson
All rights reserved
Printed in the United States of America
Published in Canada by HarperCollins*CanadaLtd*
Designed by David Bullen
Library of Congress catalog card number: 86-62837
THIRD PRINTING, 1995

North Point Press
A division of Farrar, Straus and Giroux
New York

For Harry Mason, Charles Washburn,
and Angus Wright.

Contents

Preface

Seven of these essays had their genesis as talks, seven originated as articles; four I wrote with no idea how they would be used. Another tabulation: eleven originated out of irritation, three out of joy, and four for my own clarification. There is no statistical correlation between the two tabulations. (The data indicate that I am as likely to be irritated when writing a talk as an article.) I don't apologize for the imbalance between irritation and joy. We all live in a world of wounds.

Parts of two of these essays appeared in *CoEvolution Quarterly*. One appeared in the *Journal of Soil and Water Conservation*, one in *Meeting the Expectations of the Land*, one as an essay in Terry Evans' recent book of photographs *Prairie: Images of Ground and Sky*; this essay also appeared in *East–West Journal*. Most of the essays have appeared in one form or another in the *Land Report*, a publication of The Land Institute.

The talks have been given at such places or gatherings as the annual meeting of the Saline County Kansas Farmers Union, Kansas Wesleyan, Dubuque Theological Seminary, Harvard University, the annual meeting of Kansas environmentalists hosted by the Kansas Natural Resource Council, the College of Marin, Concordia College, and at the annual meeting of the California Organic Growers Association.

None of these essays was written with the idea that it would appear in a book. They are responses to conference themes, to environmental ills, to the desperate predicament of farmers, and to the queries of graduate students, fellows, and faculty of the John F. Kennedy School of Government at Harvard University. But most importantly, they are the outgrowth of the daily work of The Land Institute; the thinking as well as the research, and the search for alternatives.

Because The Land Institute is the breeding ground for most that is written here, a word about our organization is in order. At this writing we have a small support staff. Eight scientists and as many agriculture interns are at work studying ecosystem-level agriculture. People here represent the fields of genetics, ecology, entomology, plant pathology, and the humanities. We have two hundred acres, about half of which is native prairie that has never been plowed. I suppose that we are idealistic and that our goals seem lofty. We sustain our focus by constantly measuring our work against the standards set by nature's prairie. But we also freely explore the social, political, economic, and religious implications of the way humans have to live and work the land that sustains us. These essays are the products, therefore, of the land and the people here.

Altars of
Unhewn Stone

Altars of

Unhewn Stone

Scarcely a day passes on these two hundred acres of The Land Institute near Salina in north central Kansas that I fail to stand on our bluff and look down at the Smoky Hill River. This river serves as a barometer measuring the consequences of recent weather and field conditions. In winter with little runoff from the fields, this Great Plains tributary in the basin of the Big Muddy, or Missouri, could easily pass for a clear mountain stream. Then the ripple below me, a ripple that results from the displacement of the Wellington shale, makes me feel that all is well in the countryside. But the mood of the river is mercurial. When its back is raised from a one-inch rain on fields just worked, it carries for days a sediment load of disturbing proportions toward a new home in the sea. The ripple mostly disappears then and the river's winter serenity moves far back in the mind. The river serves as a seasonal reminder of the consequences of till agriculture over the watershed.

Beyond the window where I sit at work is a field of wheat—a legacy of the Mennonites—and a patch of prairie far more native than I or the native Americans before me. It was from these two nearly opposite conditions of the land that I began to concentrate on the problem of agriculture and the possibilities for a new agriculture, one that would allow streams to run mostly clear, not just here in the heart-land, but everywhere.

In *New Roots for Agriculture* I described those possibilities, while examining what I now call the four great failures: the failures of history and prophesy, of organizations, of stewardship, and of success.

During the late 1970s I examined the history of soil abuse around the globe to learn how extensive the effort has been over the centuries to warn humanity about the consequences of erosion. I thought perhaps there were not enough people sufficiently enlightened or passionate to call adequate attention to the problem of soil erosion. But the record seems clear and, for all practical purposes, complete. Prophet and scholar alike seem forever to have lamented the consequences of wasteful farming. In the Bible, Job warned that "the waters wear the stones" and "the things that grow out of the dust of the earth" are washed away, leading to the eventual destruction of "the hope of man." Plato called attention to the "mountains in Attica which can now keep nothing but bees, but which were clothed not so very long ago, with . . . timber suitable for roofing very large buildings. . . . The annual supply of rainfall was not lost, as it is at present, through being allowed to flow over the denuded surface to the sea. . . ." The native American leader Tecumseh, in a speech full of fire and vengeance, exhorted his people to return to their primitive customs, to throw aside the plow and the loom, and to abandon the agricultural life. He warned that white men would subject them to servitude after they had possessed the greater part of their country, turned its beautiful forests into large fields, and stained their clear rivers with the washings of the soil. The eighteenth and nineteenth centuries were filled with warnings from the enlightened, impassioned, and articulate. But while we have seen improvement here and there over time, on the whole, the problem has grown worse.

In thinking about the second great failure, the failure of organizations, I wondered how the Soil Conservation Service (SCS), after nearly fifty years in existence, can have so totally failed to reduce soil loss to amounts that can be replaced annually. In fact, soil loss in the U.S. is much greater now than when the energetic and imaginative

Hugh Hammond Bennett established and staffed the SCS with men and women, from office workers to Ph.D.'s in soil science, dedicated to the common task of saving soils. Wellington Brink said that never in modern government has there been an organization with more *esprit de corps* among the workers. There was good intention at every level, an abundance of passion about the problem, and the funding for programs to carry it out. The SCS built tens of thousands of miles of terraces, prepared demonstration plots and hosted field days. One two-day program held on two adjoining farms in southern Ohio attracted 80,000 people! And the SCS had help from private organizations—for fifteen years a strong citizens' group worked alongside it. They called their private organization Friends of the Land, published an attractive *Land Quarterly*, and advertised themselves as a society devoted to the conservation of soil, rain, and man. They disbanded in the mid-1950s, but during their active tenure, nearly all the great names in conservation contributed to the organization and the quarterly. Here was an organization that failed to sustain itself, let alone solve the problem. The sad verdict is that public and private organizations have failed to stop soil erosion.

Though I count the efforts of the SCS to be a failure, I do not mean it should be scrapped. In fact, it should be enlarged, as a matter of national defense. Imagine how much the nation's soils would have deteriorated without such an effort.

I identified the third failure, the failure of stewardship, after I had seen soil erosion even on Mennonite and Amish fields. This was particularly disappointing, for the people of these closely related religions believe there is no higher calling from God than to farm and to be good stewards of the land. They take seriously the biblical injunction to "dress and keep the earth." Within the context of till agriculture, they are the most ecologically correct of all major groups of farmers in the U.S. Observing erosion on their lands, I have wondered if it may be beyond our ability to achieve a sustainable agriculture century after century, if till agriculture is its main feature.

Success may be the worst of the four failures, simply because we learn very little when all goes well. It has been argued that farmers can't be expected to take care of their land without good prices. While there is truth in that argument, high prices have not guaranteed wise farming. During the few periods when prices *have* been good, erosion has remained a serious problem. When Secretary of Agriculture Earl Butz told farmers to plow fencerow to fencerow in anticipation of a huge export market, they did, with the result that hedgerows and shelter belts were bulldozed. The worst part of this failure is that high yields give us the illusion that all is well on the farm: even as soil is being lost, yields climb; as pesticides and fertilizer accumulate in the soil and groundwater, yields climb; as the fossil-fuel energy cost per acre of production escalates, yields in every major crop go up. Even though low prices due to overproduction have contributed to the modern debt crisis on the farm, production remains the bottom line. Unfortunately, success in producing food tends to blur the cost in ecological deterioration and in the depletion of fossil energy resources. High-production yields numb us to the understanding that all of these costs must eventually be reckoned with.

Alfred North Whitehead wrote that the essence of dramatic tragedy lies not in unhappiness, but in the remorselessly *inevitable* working of things. As agriculture stands now, and as it has stood more or less for the last ten thousand years, it seems to fulfill Whitehead's definition of tragedy. We need food and, given the current human population now dependent on till agriculture, we will need to continue to till the soil, even though such activity has historically and prehistorically undercut the very basis of our existence. We live in a fallen world.

The Fall, at least as Christians have understood it from Hebrew mythology, can be understood in a modern sense as an event that moved us from our original gathering-hunting state, in which nature provided for us exclusively, to an agricultural state, in which we took a larger measure of control over our food production, changing the face of the earth along the way. Because we live in and depend so exclusively on till agriculture in this "fallen world," we could easily shrug our shoulders and live our lives as though there were little we can do to change things. But none of the great religious thinkers that I know accept shoulder-shrugging over the Fall as the final response. Even for people who insist that religion has no claim on them, the "gift of denial" is part of human nature. Citing Freud, Garrett Hardin wrote years ago that no one believes in his own death. People who ride motorcycles without a helmet may be aware of the risk of death but deny that it will happen to them. Yet the gift of denial also allows us to act positively and to think creatively. It can supplement what amounts to a religious view that we must do our best to prevent further deterioration of this fallen world; indeed, that we must restore it.

Around where I live and work, I have found it easy to deny these troubles. While soil erodes in the rolling wheat fields, it stays put in pastures and in native prairie grassland. Soils have stayed put for a long time on the prairies, *independent of human action*. With the wheat field come pesticides, fertilizer, fossil energy, and soil erosion. The prairie counts on species diversity and genetic diversity within species to avoid epidemics of insects and pathogens. The prairie sponsors its own fertility, runs on sunlight, and actually accumulates ecological capital, accumulates soil. Observing this, I formulated the question: Is it possible to build an agriculture based on the prairie as its standard or model? I saw a sharp contrast between the major features of the wheat field and the major features of the prairie. The wheat field features annuals in monoculture; the prairie features perennials in polyculture, or mixtures. Since all of our high-yielding crops are annuals or treated as such, crucial questions must be answered: Can perennialism and high yield go together? And if so, can a polyculture or mixture of perennials outyield a monoculture of perennials? Can such an ecosystem sponsor its own fertility? Is it realistic to think we can manage such complexity to avoid the problem of pests out-competing us? Since raising these four key questions, we at the Land Institute have devoted all of our research to answering them.

When individuals pay attention to the problems of soil erosion, toxic chemical pollution of our soils and water, salination of irrigated soils, and the dependency of agriculture on a finite supply of fossil fuel, *then they inevitably base their concern about the health of agriculture more on long-term sustainability than on short-run production.* At this point the conflict between humans and nature becomes apparent, for agriculture is primarily production oriented, while nature's emphasis is upon preserving potential. For nature, production need only be sufficient in order to ensure that *potential* is preserved. Humans reward enterprise, while nature rewards patience. For example, a one-acre Illinois cornfield will contribute several times more weight gain to a feedlot steer than the acre of Illinois prairie that once stood in its place. Nevertheless, those same Illinois soils now being mined by highly bred corn varieties were built by the prairie sod that we plowed under. Early in this century, Sir Albert Howard used the forest as an analogue in thinking about and experimenting with different approaches to food production. At the Land Institute, we are using the prairie.

Shortly after the publication of *New Roots for Agriculture*, we began to assemble a small research staff and to enroll ten agricultural interns, mostly graduate students, a year, to attend class and do research during a forty-three-week term, from mid-February through mid-December. Both staff and students poked around for relevant sources of information useful for the kind of agriculture we envisioned. After a while, we came to rely more and more on a body of knowledge largely ignored by most agricultural researchers. It was knowledge that had been accumulating in the synthetic fields of evolutionary biology and population biology, in population genetics and ecology. This knowledge was of an essentially pristine nature, in that it had been accumulating on the shelf more or less for its own sake. But it seemed essential for the kind of agriculture we envisioned. As I began to talk to scientists in various universities across the country, one striking fact became apparent: the scientists responsible for discovering this pristine knowledge were not aware of its potential for solving any of the ecological problems in agriculture.

It became clear to me that there is more to be discovered than invented, and that the future of the human species on this planet will depend more on discovery than on invention. Long ago, Charles Lindbergh said, "The future of the human race will depend on combining the cleverness of science with the wisdom of nature." But through domestication we have removed our major crops so far from their original context that most farmers (and I suspect, most agricultural researchers) regard both crops and livestock as more the property of humans than relatives of wild things. Yet all of these creatures evolved in ecosystems that had little to do with humans. None of these ecosystems was of our design. Somewhere in the midst of this thinking and research the scripture in Exodus 20:25, from which the title of this book came, took on a deeper meaning for us. Right after Moses had delivered the Ten Commandments, he received instructions to build an altar of unhewn stone "for if thou lift up thy tool upon it, thou hast polluted it."

This scripture must mean that we are to be more mindful of the creation, more mindful of the original materials of the universe than of the artist. The altar was to stand as a reminder that we could not improve on the timeless purpose of the original material. I don't think such a scripture means that we are *never* to shape the earth with our art or our science, but that the scientist and the artist must remain subordinate to the larger Creation. The chances of disrupting nature's patterns, upon which we are dependent, are greatly reduced if we assume this modest posture.

While translating our thinking and research on our particular concerns into priorities for more reading, thinking, and experimental design, we could not ignore the immediate crisis on the farm because of the industrialization of agriculture. Many of our neighbors were going broke. Thousands of people across the land were going broke, losing land that had been in their families three and four generations and more. We have been unable to help the farmers in the here and now. They need solutions to their problems fast. But observing their difficulties, we could not help but think about the farmers who have managed to remain financially solvent and about the combination of factors that has made them so, including the ecology of their farms and their relationship to what is left of the surrounding rural community. In other words, natural ecology, the ecology of modern farms—from worst to best—and the ecology of a future agriculture, have all been rolled into one continuous thought process.

Wendell Berry has said that when we have destroyed the forests and prairies to replace them with agriculture we have never known what we were doing because we have never known what we were *un*doing. By studying the natural ecology we have a chance to see what we have *un*done. We can compare our modern farms, from worst to best, with what remains of our natural ecosystems. By making these close-at-hand comparisons of what we have done during our period of *un*doing, we can think more clearly about the ecology of a future agriculture.

These essays, then, are an attempt to understand the requirements of a science to be pursued as though its original material is more important than the work of the scientists who are shaping that material. What if we researched and taught as though we believed that the wisdom of nature *is* more important, in the long run, than the cleverness of science? What if we really did regard our domestic plants and animals more as the relatives of wild things than as our property? What if we acknowledged straight out that there is more to be discovered than invented? Of course we must have both discovery and invention, but what if we changed the emphasis?

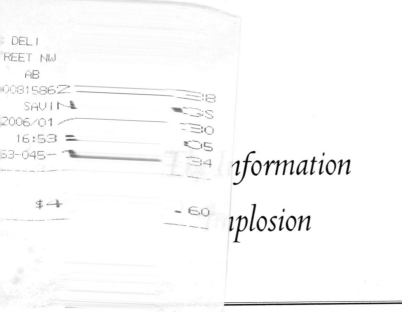

Information
Explosion

Though conventional wisdom holds that we are in the midst of an information explosion, more careful consideration must surely convince us that the opposite is true. Think of all that has happened to the world since 1935. Few dispute that there is less biological information. Species extinction at the rate of one thousand species a year or so, especially in the tropics, coupled with the genetic truncation of the major crops, undeniably is a major loss of biological information. The new varieties released by plant breeders do not represent more information. "Variety" is a legal term and reflects a selection of biological information already present. Selection of a single gene for an entire plant population can result in a named variety even though that "variety" represents only a single gene at a single locus responsible for a particular virtue that a seed house wants to peddle.

Species extinction and genetic narrowing of the major crops aside, the loss of cultural information due to the de-population of our rural areas is far greater than all the information accumulated by science and technology in the same period. Farm families who practiced the traditions associated with planting, tending, harvesting, and storing the produce of the agricultural landscape gathered information, much of it unconsciously, from the time they were infants: in the farm household, in the farm community, and in the barns and fields. They heard and told stories about relatives and community members who did something funny or were caught in some kind of tragedy. From these stories they learned basic lessons of agronomy. But there was more. There was the information carried by a farmer who looked to the sky and then to the blowing trees or grasses and made a quick decision as to whether or not to make two more rounds before quitting to do chores. Much of that information has already disappeared and continues to disappear as farmers leave the land. It is the kind of information that has been hard won over the millennia, from the time agriculture began. It is valuable because much of it is tuned to the harvest of contemporary sunlight, the kind of information we need now and in the future on the land.

A friend of mine, a distinguished professor in a major university, is terribly alarmed about species extinction in the tropics. He is a leader in the fight to save rain forests everywhere. As a person who has joined the fight to preserve the biota of the planet, he gives numerous talks each year about the problems of overpopulation, resource depletion, and pollution. He heads the library committee for his university and is much impressed with the "knowledge explosion," how much we now know, and how much better educated graduate students are now than they were when he was a student. As do most Americans, he sees Silicon Valley and the computer industry as representing an expansion of knowledge. When I suggested that there is less total cultural information in the U.S. today than fifty years ago, he did not agree. I was thinking about the cultural information just mentioned, the information that has left the countryside, the kind of information that is a necessary basis for a sustainable or sunshine agriculture. What we had in 1935 was, of course, not adequate for saving the soils or preventing the countryside from being chemically contaminated. There was too much rural ignorance, cruelty, and xenophobia, and no one would want to romanticize that side of humanity or culture.

Nevertheless, where we were then was a better takeoff point for where we need to be than where we are now. There was more resilience in the culture at large because more people were on the land.

And though in rural areas it was a "cream and eggs economy" more than a "money economy," most people in the cities had relatives on farms and therefore had this second economy to fall back on during hard times. Because the fabric of community was more intact, ordinary human problems were dealt with more directly, inside the community rather than outside it. Under the watchful eyes of adults, rural teenagers experienced rites of passage, such as putting up hay or canning tomatoes. With those rites of passage largely gone, it's little wonder that teenage drug and alcohol abuse and teenage suicide are so high.

My concern here is the serious reduction of people on the land who can pass on to future generations the skills, the traditions, the passions, and the values they will need to farm well on the smaller energy inputs inherent in the use of "contemporary" energy. Contemporary energy is that which arrives from the sun and is harvested in a horizontal manner over the landscape, rather than from a vertical well or mine. First this energy is trapped by plants, then it is harvested by humans. Today we are dependent upon old, or fossil, energy, extracted vertically through mine shafts, strip mines, and wells. Although contemporary energy has a low density per square foot, its supply is ensured. But it requires high cultural information to harvest and store it safely for future human use. When the density of a mineral, for example, is low, more thoughts and combinations of thoughts are necessary to make it available. The same is true of low-density energy. More cultural arrangements are required to harvest a given quality of sunlight through solar collectors than to make the same quantity of energy available by burning coal at a power plant. The information now piled up by scientific discovery of how the world works pales by comparison. Even though most of this scientific knowledge is likely to stand after the fossil fuel glut is over, the technological array put in place to exploit this knowledge is less trustworthy, for it will have occurred in a fossil fuel–based infrastructure and will likely not be appropriate for a sun-powered future.

The culture believes that we are in the midst of an information *explosion* because of the status granted the knowledge accumulated through formal scientific methods. In contrast, knowledge accumulated through tradition, daily experience, and stories, mostly in an informal setting, has little status. We have taken this "folk knowledge" for granted, I suspect, for however *complex* it might be, it was not all that *complicated* to internalize. What we acquired second nature was woven in with the rural setting, the daily work, the local values and moral code. It is more the legacy of the dead than of the living. The more respected body of knowledge, learned through formal discovery or revelation of discovery in classrooms and textbooks, is of a different order. More discipline is involved both in the discovery and in learning about the discovery. And though most of this information is not all that complex, it is more complicated for us to learn and internalize. Maybe this is the reason we assign greater value to such knowledge than to that which we picked up through tradition. There has been an explosion of formal knowledge, but what was necessary to make it accumulate so fast led to the destruction of much of the other older, less formal knowledge.

Spread across the land surface of the planet, tuned to local environments, with potential to renew the earth and run on sunlight, species and individual organisms are special creations for the spaces they inhabit. The loss of such diversity from the landscape is very serious. Like my professor friend, I worry about this loss of genetic stock, for it is a loss of the most important form of information on the planet. But the loss of cultural diversity across the land surface, cultural diversity that was just beginning to be more tuned to the local environments of our recently discovered America, is also serious. I suspect that we pay this disappearing diversity such little respect because of the *illusion* that knowledge overall is more plentiful. Species diversity has been hard won. Numerous deaths stand in the background, in the evolution of the current life on earth. Cultural information, including agricultural information, has been hard won, too. Countless deaths stand behind this information, as does lots of anguish and hurt. That is why rural places have traditionally been the source of the *lasting* values of a culture. What my professor friend and most of his allies have not grasped is that the war against the tropics is the same war that is being waged against agriculture and rural culture.

Part of that war against rural culture can be seen in the negative attitudes of our larger culture toward rural places and rural people. They run as deep as the worst forms of racism. A reviewer of the film *Country* said that Jessica Lange was too beautiful to be a farm woman. A reviewer of a recent book by Wendell Berry said that although Berry was a farmer, he was "an intelligent farmer." People who would be outraged if they heard a black called "nigger," or a woman a "little girl," make such statements about farmers and see nothing wrong with them.

A biologist friend at another major university, who is concerned about species extinction, revealed his prejudice against rural places and rural people. During an exchange of family gossip, I mentioned that our son was a student in Lawrence, at the University of Kansas. He said he was sorry my son was in Lawrence. I said that he liked it there, to which my friend replied, "Well, I guess it is better than Salina." What I suspect was at work in the mind of this professor was a combination of cultural snobbery at and boredom with what he considered the unglamorous Kansas landscape. Salina, Kansas, is a town of forty thousand people, most of them of rural origin, descendents of those who possessed information about the farms and ranches on which they had been raised and from which they eked out a living. They may have done stupid things on those places, and many were probably poor at their work, which made them lose their farms. But most of them were driven from the land by the industrialization of agriculture.

Their experience exemplifies a law at work in the world, a law of human ecology: high energy destroys information. High energy (such as fossil fuel or nuclear energy) contributes to the arrogance of university professors who, though righteously appalled when species disappear, pay little attention to farmers driven from their lands or to the loss of cultural information this represents. This cultural information, which was hard won through sweat, tears, injuries, and death, will have to be won back in the same manner, and not just for the land, but for the urban culture too. Though cultural information can evolve faster than biological information, once lost it will be difficult to regain. Reestablishment will be gut-wrenching and the land will experience further abuse. The eyes-to-acres ratio will have been even more distorted, and I fear that the industrial model for agriculture may be regarded as even more necessary in the last years before the collapse.

Old Salsola

Ten A.M., January 2, 1986. Through the windows of John's cabin, perched on a low bluff, I look northward over the Cheyenne Bottoms near the Great Bend of the Arkansas River. A strong wind blows from the northwest. Between where I sit and the expanse of water in the bottoms lies a poor pasture, some alfalfa fields, and newly worked sorghum ground. Tumbleweeds of the genus *Salsola* roll across these fields like purposeful animals migrating to some destination beyond the horizon. They *look* purposeful anyhow. They look at least as purposeful as the skein of geese that rises out of the bottoms each morning, forming long poor V's and heading south to shop for breakfast in the fields of shattered sorghum. The wind that propels these weeds and the wingbeat that propels the geese are from the same source: transformed sunlight. Barbed wire fences have stopped some thousands of these weeds, but thousands more roll right over, *even* where their dead relatives have not accumulated against the wire. A moment ago, one particularly bouncy weed rolled right over a fence, bounding almost like a deer, but with one important difference: a dead *Salsola* mother will hug the same wire that a live deer clears. Free of the fence so briefly hugged, the cheerful dance of the dead continues. I wonder what is the average number of seeds dropped at each bounce. Surely less than one, but there are lots of bounces in this winter trip of a dead *Salsola* dispersing her children.

Was it she or the "larger system," an ancient ecosystem, that prepared her for this day of winter wind: Born in June, her branching pattern makes her round by fall, testimony to her ability to remember the past and foresee the coming season with each cell division. But there is more to this globular weed than her shape. For all through the summer, at the base of her stem, she formed an abscission, a knotted ring of cells for easy detachment at ground level in late fall. I don't know whether she or the larger ecosystem was most responsible for that knotted ring, but the wind does seem a fitting hearse for a last ride to a fenceline cemetery. What other plant could beat *Salsola* in this respect: that it is only in her death that the most energetic and widespread dispersal of her offspring could happen?

Biotechnology and
Supply-Side Thinking

Genetic variation is the consequence of mutation and selection by nature over hundreds of millions of years of evolution. Biologists regard the first agriculturists as the foremost revolutionaries of all time, not because they gave us nearly all of our crops and livestock, but because they were the first to short-circuit the processes of genetic selection that had been at work in nature since life began. The age of modern plant breeding, in which we employ powerful statistical tools supporting highly sophisticated experimental designs, is not a new age so much as an extension of the work done by the founding mothers and fathers of agriculture.

Now we are in the age of biotechnology, an age abounding with promises that new technologies will allow us to short-circuit even further the ancient long-term processes of genetic selection. Though biotechnology is a means of short-circuiting nature, it still ranks lower in importance than humanity's fundamental split with nature some ten to eighteen thousand years ago. That split was profound, for when we took charge of our food and fiber production and thereby transformed the landscapes that shaped and sponsored us, we placed ourselves on a treadmill that now requires eternal vigilance in the procurement and application of knowledge unknown to any other species.

My opposition to many of the grand claims of biotechnology is not because they are new, but because they are painfully old. They are part of the recent acceleration of the fall that came with the industrialization and chemicalization of agriculture. The ancient split between humans and nature widened during this epoch. This recent flurry of interest in biotechnological techniques has its roots in the same motives that fueled the transition to industrial and chemical agriculture.

There is one new possibility in the biotechnology grab bag—our ability to transplant genes from one organism to another remotely related creature. Little else in that bag is really new. Cloning is old and easy, as anyone who has had anything to do with a strawberry patch knows. Tissue-culture techniques allow us to cut corners considerably, but we should remember that cloning is simply monoculture in spades. Biotechnology involves genetic manipulation. Though it involves biochemistry, it is still genetics, itself a young discipline. Perhaps because the field of genetics is still young—only as old as this century—we have yet to develop a mature perspective on its promises and pitfalls, or a calm language.

Let us look at some history. In the 1930s genetic mapping of chromosomes became possible. It became useful, too, for us to know what genes were linked together on the same chromosome and how far apart they were. Mapping helped breeders calculate how many plants they would have to grow out and the acreage needed to recover certain desirable recombinations. After the techniques for genetic mapping were elucidated, geneticists made enthusiastic claims that they would eventually map all genes in humans, corn, fruit flies, mice, and the like. What happened, of course, is that they mapped the easy five or ten percent and then gave up, deciding it wasn't worth it.

About the same time that genetic mapping became prominent, certain geneticists made a big fuss over the discovery of polyploidy in plants. Polyploids are organisms, mostly plants, that have more than two sets of chromosomes. Early on, geneticists believed that gigantism accompanied polyploidy. They talked of growing big plants in much the same manner as they talk today about breeding cows as large as elephants. As yet, I can't think of a single crop in our inventory that is the consequence of human-induced polyploidy. There may be a few insignificant horticultural forms, but results did not match the early enthusiasm.

But this wasn't the last heyday in genetics. After scientists learned that radiation could induce mutation, they and the press were even more enthusiastic about the new tomorrow that was to visit the crop world. Seeds were stuck in atomic piles for prescribed periods, removed, planted, and observed from germination through seed set, for several generations. Several scientists spent most of their professional life on such studies. Professor Walter Gregory's work with the peanut was probably the most extensive, covering several decades. What do we have to show today? Very little. The reality is that breeders soon realized that they had the same problem with radiation-induced mutants that they had with wide hybrids—the difficulty of handling all of the genetic variation coming at them. Radiation-induced varieties as a goal for agronomists have been essentially sidelined.

Many of us experience *déjà vu* when we hear or read of the promises of biotechnology. I think of genetic maps five percent complete, of polyploid research that died when a primary investigator retired, radiation genetics that essentially provided the early core for the field of biophysics. I think of all the plastic and brass labels that have been quietly removed from laboratory doors in biology departments across the land, leaving embarrassing holes where screws once supported markers of those proud little empires. Meanwhile, the promoters, both in genetics and in the press, seem to have disappeared into the woodwork—after the money has been spent. Some of the discoveries have been useful, but what if the money had been spent on other social goals, even on research in genetics?

Two examples of underfunded yet socially desirable research come immediately to mind. The Opaque-2 mutant in corn is high in the important amino acid lysine. Opaque-2 would make corn a higher protein crop. Yields would be somewhat lower, but grain quality would be higher. Very little money was spent in promoting this mutant and even less on research to optimize its genetic background and make it an even more desirable plant food. Apparently, the Chicago Board of Trade, accustomed to handling quantities of a known carbohydrate producer, could not handle the high-quality corn in its inventory. But if we had the political will to require the Board of Trade and its allies to handle high-lysine corn, it could become an important crop.

The history of work on the quantitative gene perhaps presents a still better example of our failure to implement an important product of science. Both the mathematical theory and the testing of the role of quantitative inheritance was done in the 1960s at North Carolina State University. Researchers there concluded that quantitative inheritance could match the yields to come from hybrid vigor. Hybrid corn could be replaced by high-yielding, open-pollinated varieties that produce seeds capable of germination without the intervention of agribusiness and agronomy. If the nation had picked this as a social goal, farmers could be planting their own seed corn nearly every year. That might be economically disastrous for the seedhouses, whose financial backbone is seed corn, but the economy of the farm itself would have been tremendously helped. Yet how many know of this work and of its potential for cutting farmers' input costs?

This brings us to a more important question. Why is industry interested in biotechnology? Because it will make money and create an even greater industrial dependency. Corporate America, in general, is interested in the simple but large effect. Corporations endorse this motivation, I suspect, because we are a magic bullet nation. We want that magic bullet to solve complex problems in one shot. This is the reason for our failure to promote both high-lysine corn and the quantitative gene. Now we are faced with the new field of biotechnology, where heroic notions of the industrial age are fueled with the desire to make money regardless of farmers' needs.

How do we combat this form of corporate control? At one level, I suppose, in much the same way that we organized for civil rights or against the Vietnam War or nuclear power. We need wide and deep public involvement on this issue. I know that some will argue that since most of the promises of biotechnology are going to fail, why not let them run their course? I see two reasons why we need public involvement: First, government money is clearly our money. If spent on an inappropriate national priority, it represents resources robbed from better social goals. Second, corporate money spent on these projects also hurts society. A waste of capital and of human resources in corporations ultimately hurts society as much as the waste of public money. The private and public economies are so inextricably interwoven that to suggest that the loss of capital in the private sector is benign is to oversimplify the problem.

Probably the first order of business is to get ourselves informed on the subject in much the same way that people who were not nuclear physicists informed themselves about nuclear power. This means we will not turn biotechnology over to a priesthood of experts. At every step, the public should challenge the underlying assumptions of those who promote biotechnological techniques. We will hear the argument that to thwart the proliferation of this technology is to "roll back science" or to "dull its cutting edge." We need to remember that lots of "cutting edges" in science offer the possibility of contributions every bit as profound as those offered by biotechnology. The fundamental question is not "What can we do?" but rather "What kind of a world do we want and what will nature require of us?" I expect that biotechnology will have a role, but it should not dominate our thinking or our budgets.

As a way of clearing our minds, let us look at some of the problems in American agriculture now. Rural communities continue to be destroyed and farmers are going broke. Along with soil erosion and chemical contamination of the countryside, we are experiencing the loss of genetic diversity in our major crops. Successful plant breeding tends to narrow the genetic base of any crop roughly in proportion to its success. The better adapted to the human purpose a population is, the narrower its genetic base and, therefore, the fewer options it has available to meet future threats of insects and pathogens. At this writing there are seventy million acres of corn in the

U.S. Thirty percent of that crop comes from four inbred lines—the same as crossing two heterozygous individuals. In the last fifty to seventy years we have set the stage for increased unadaptability to unforeseen and unforeseeable crises.

At two conferences at which I was a panelist, audiences raised the question of whether the gene splicers have arrived just in time to reverse this genetic drain on our major crops. But attempting to introduce a foreign gene into a new crop may turn out to be more formidable than we might think. The gene splicers may be successful with the easy tasks—but inserting an ice-nucleating gene into a bacterium is simple compared to accomplishing a similar trick in a higher organism. And gene splicers must still solve the problem of optimizing a gene against a larger genetic background that is alien. In any individual specimen, all genes interact. The constellation of that individual's native genes must tolerate the introduced gene. But even if the splice works in an individual case, there remains the problem of reoptimizing that gene at the larger population level. There are episodal genes to be reckoned with—genes that may cause a hormone to be released at a specific time, generating major effects on development. Since all genes interact, we can't simply throw them into a recipient plant container to swish around at will.

There is no guarantee that the new techniques will do anything more than accelerate the narrowing of the genetic base. To be able to transfer a few genes or even several hundred genes is not what we need. We need to ensure a steady flow of genes from well-adapted reserves where optimization has already occurred. The need to maintain these reserves will become greater, not less, for with selection we trim the genetic base, from which we must replenish the gene pool. Therefore, so long as we practice plant breeding we will need new stocks to meet the demands for local adaptation. The increase of the human population will provide a strong incentive to push our crops into new environments. New diseases will arise; social demands will change. The green revolution in wheat and rice should have taught us the seriousness of breeding high-yielding varieties thousands of miles from where they eventually displaced the numerous local varieties and mixtures. The new varieties were well adapted to high-input agriculture, but the genetic variability of the new varieties was less. The total genetic variation over those landscapes went into steep decline.

The upshot of all this is that some of the capital going into biotechnology would be better spent on genetic conservation in order to compensate for the decline in native variability, but doing that would not make a biotechnology firm rich.

And now I want to consider the other end of the spectrum. If biotechnology can be characterized as relying primarily on human cleverness and less on nature's wisdom, applications from the science of ecology are just the opposite; they rely more on nature's wisdom and less on human cleverness.

It is still important to ask how organisms carrying new or altered genes will relate to the ecological interrelationships of organisms when introduced into the environment. Based on what is known about mutations responsible for major changes in either the physiology or morphology of an individual, I suspect that the big changes won't haunt us so much as the small cumulative changes that we will encounter after we are deep into the new era. In other words, the peculiar-looking genetic monster is less likely to be adaptive and therefore is less of a threat than the slow incremental changes that push us toward ever-increasing homogeneity of germ plasm.

The early agricultural revolutionaries surely did not *intend* to give us the problem of soil loss, yet more carbon is in the atmosphere due to till agriculture than from the industrial revolution. No one intended this. Those who gave us electricity from coal-fired power plants did not intend to give us acid rain either.

We can understand the life support system of the spacecraft because we built it out of mechanical parts, but as Eugene Odum says, "the biosphere is bioregenerative and self-regulating. Since we did not build it we don't know much about how it really works, and we have shown little interest in studying it at the necessary large scale until recently, when malfunctions have begun to appear due to human impacts. Because microorganisms play major roles in maintaining earth's life support systems, we need to be especially careful about tinkering with decomposition and other recycling processes." (*Science*, vol. 229 [1985]: 1338).

I read recently in *Science* about our problems in cleaning up industry. Those working on the ecological impact of industry have come to realize that though there is still a great deal of ecological knowledge to be gained, we tend to use less information than is already available. The point is that we need to do a better job at employing the information we have.

Once we said that we did not know enough. Now we are saying we know a lot, but we don't employ our knowledge. The implication is that we need to *apply* what we know. But developing from these considerations is the likelihood that *any system we use for organizing the information we have gathered also has limits* and that humans will never be able to organize the necessary information we have on the shelf in a way that will serve to protect a biosphere that has been overly tinkered with.

As some of us noted the alarming rate at which fossil fuels were being burned, we began to think about how to change our minds from regarding them as "burn it up" fuels to regarding them as transition fuels—fuels to be used in making the transition to a solar-powered future. We promoted the solar technologies (wind, hydroelectricity, biomass conversion) and tried to cut our fossil fuel consumption. We did not assume that we should find a new source of energy that would further supply our binge of energy use. Instead we looked for sustainable energy sources.

Soil is as much a nonrenewable resource as the fossil fuels. In its use we are forced to the same question: How can we use our soils in such a manner that we can make them last until sensible agricultural ecosystems are moved onto the landscape?

Finally, we have to think of the genetic narrowing of our major crops and consider a similar, sobering question: How do we wisely spend the declining genetic variation in our major crops during this period of transition from an extractive to a sustainable economy? With the spirit of colonization in our bones, we have been preoccupied with new sources, with little time to consider conservation. Consequently, we look for new sources of high energy—nuclear energy, for example—to use when the coal and oil are gone. As soil erodes and landscapes have to be abandoned, we look elsewhere to "open up" new lands, meanwhile giving insufficient attention to conserving what we have. Now we are at war against genetic diversity, both in species and in our major crops. With each cut of selection the breeder eliminates parts of the genetic code from the inventory, parts necessary for a healthful and productive future. Just as nuclear technologists have told us not to worry about losing the fossil fuels, enthusiasts for biotechnology are making us think that they can transfer genetic information into our major crops from wild populations and that they can splice in these genes from this abundant source for a long time to come. Such a technology feeds the colonizing mind. But colonization is based on plunder and is inherently violent and wasteful, a fact we have been slow to acknowledge. The antidote to colonization is discovery. We need to discover how the world works to know better our place in it. In this sense, the true discovery of America lies before us. So far we have mostly colonized it.

Scientific Balloons

In the last essay I mentioned some of the promises made in genetics over the past fifty years—complete chromosome maps, gigantism in plants due to polyploidy, radiation-induced mutants that were to speed up our breeding programs by providing a new source of genetic variation. Very few of these promises have materialized. Now we are hearing of the bold promises of biotechnology—wonder cows and hogs, wonder crops, and so on. I recall as well the promises, made thirty years ago, that nuclear fission was to bring us energy too cheap to meter. So far it has brought us high utility bills, the scare at Three Mile Island, and the serious disaster at Chernobyl. Now we have to worry when promises do materialize.

It is a virtue to forgive and forget a stupid or wicked act. But our ability to forget human folly before we are required to forgive is another matter. If people do something stupid or wicked and do not acknowledge it, we count it legitimate to call the mistake to their attention. We hope and expect to see some sign of contrition and, more importantly, a lasting change in behavior.

Nearly everywhere, the scientist enjoys the trust and respect of the population. This high regard has allowed scientists, as experts, to work at such a distance from the culture at large that few of their discoveries and resulting technologies have fallen under careful public scrutiny. Consequently, numerous discoveries or partial discoveries entice both scientists and the press into wide-ranging speculation about their possibilities and implications. The uncertainties or apprehensions of the scientists who gathered the data are seldom mentioned. Often to their own surprise, thousands of scientists and technologists find themselves rising in a scientific balloon, on air made hot by scientific fame-seekers eager to talk to the press. Most of the balloons fall and are forgotten. The scientists who rode them and fueled them are forgotten, too. Yet no one calls for an accounting. No one acknowledges the cost to society. No one questions to what extent the effort was motivated by the desire for power and fame. No one is embarrassed. No one asks forgiveness of the taxpayers who funded these failures, and the culture forgets before there has been a chance to forgive.

Building a

Sustainable Society

In the August 1984 issue of *Contact*, newly appointed Kansas Wesleyan president Marshall P. Stanton wrote: "The first work task given to our foreparents was to care for the Garden of Eden. Now as we view the accumulation of waste products from an industrial civilization, a haunting question comes: Is it too late to care for the Garden? The early disciples of Christ were stirred to serve faithfully as stewards of the 'mysteries of God' [1 Cor. 4:1]. Nowadays, Christians are to faithfully manage their financial, temporal, and personal talents/treasures. We are all held accountable *daily* for such stewardship."

"The first work task given to our foreparents was to care for the Garden of Eden." As a biologist whose daily work centers around my training in evolutionary biology, I am not at all bothered by the metaphor of the garden that the ancient Hebrews used. It is an accurate way of describing our relationship to nature. It is consistent with another Biblical injunction that we are to "dress and keep" the earth, an insight more complete than the insight of Darwin, who discovered the secret of biological diversity, or Einstein, who discovered the secret of the universe itself and expressed it in his famous equation $E = mc^2$. These discoveries represent important moments in our history as a species, but the ancient Hebrew or group of Hebrews who received the revelation, from whatever source, that our *first charge* is to caie for the garden, had a revealed truth of more importance and lasting significance than the combined discoveries of Darwin, Einstein, Galileo, Copernicus, Newton, and all the rest of the scientific community. This first commandment—to care for the garden, to dress and keep the earth—explicitly tells us how we are to live. It is unequivocal, unambiguous. It cannot be corrupted by thousands of interpretations as to what it means, interpretations that would serve the lusts, the desires, the greed, and the envy of mortals. It is an injunction put there, right up front, at the head of the scriptures, for good reason. It says implicitly that we are not to risk the destruction of the garden for the sake of political ideology. We are not to risk the destruction of the garden out of fear of attack from others. We are not to risk the destruction of the garden with an economic system based on greed and envy. We have one job to do, single-mindedly—to care for the *Creation* that the ancient God of the Hebrews and our own God found to be good and loved.

I suppose this means that we are not to risk destroying the garden with nuclear weapons. But we seem poised to do just that. Senator Mark Hatfield said, "To continue this nuclear madness is to shake our fist in the face of God and say, 'We, a nation, have the right to destroy your creation!' We are challenging the Creator. It is a rebellion against God." This is a U.S. senator talking, not some religious fanatic.

Never has there been such absolute authority by so few over so many. Think of it. The index finger of one man could destroy in a day a creation that has been under way for the past four billion years. No monarch in history could do that. The power at the fingertips of so few is beyond comprehension. If we took the total stockpile of nuclear weapons and converted it to Hiroshima-sized bombs, we would have to detonate one hundred bombs *per day* for *thirty-nine years* before we exhausted our supply. And now with the new star wars proposal, we are taking these weapons to the heavens, heightening the probability of the nuclear winter.

I am sure that most readers are familiar with the recent studies that pertain to nuclear winter. First there is the computerized study done by a group, headed by Carl Sagan. This study is now well-known among climatologists, geophysicists, and biologists here and abroad. Briefly, the model demonstrates that a nuclear war involving the exchange of only a fraction of the total American and Russian bombs would change the climate of the Northern Hemisphere to a nuclear winter. Soot and dust from the explosions would rise to block sunlight. Later, ultraviolet light would blind land animals because the atmospheric ozone that protects the earth from harmful irradiation would be destroyed. Radioactive fallout would be higher, over land areas larger than was once predicted.

The second study was a followup by twenty distinguished biologists, headed by Paul Ehrlich. This study predicts the extinction of much of the entire biosphere, including the southern hemisphere, as a result of the nuclear winter. What is amazing is that the report of the twenty biologists represents a consensus of about forty biologists at a meeting in Cambridge, Massachusetts, in the spring of 1983. Scientific meetings are not known for much consensus.

What these two studies illustrate is that the rules are now different. The damaged parties are no longer just the nations involved in the exchange. Earth is the loser. Soviet scientists agree with the conclusion of the Americans; both groups of scientists talked back and forth by video phone. The gory truth is that nearly half the earth's population would die very soon after a nuclear holocaust. About a billion would die immediately, from radioactive fallout and from disruption of life-support systems. Eventually, all 4.5 billion humans would probably die, for what we call nature would probably experience a mortal or near-mortal blow. Our planet can well be bombed back to supporting the level of life forms here a billion years ago, when creatures like today's bacteria were the dominant life forms.

The rules *are* different in another way, for nuclear warfare is no longer a political matter. It is a problem for us all. Scientists are going to have to inform their governments. Journalists have to inform the citizens of the world, as Lewis Thomas says, "in detail and over and over again, about the risks that lie ahead."

The Soviets now announce that as much or more radiation was

released at Chernobyl as was released in the atomic bomb attacks on Hiroshima and Nagasaki. Possibly twenty-four thousand people are expected to die as the result of the accident. The use of the "peaceful atom" has given us a first-hand insight into the consequences of a nuclear war. Soviet Premier Gorbachev told Soviet viewers that the tragedy at Chernobyl provides us with a warning about the arms race. According to Gorbachev, "Experts have estimated that the explosion of even the smallest nuclear warhead is equal to the radioactivity of three Chernobyls. If that is so, the explosion of even a small part of the existing nuclear arsenal will become an irreversible catastrophe."

There is a time factor to keep in mind. If we manage to make it through the next thirty to forty years, essentially all of the proven reserves of oil will be gone. As recently as 1974, with the then-exponential increase in coal consumption, we had a 111-year supply. But such a supply of coal will last about thirty to forty years, if energy consumption continues along an exponential path. From this we can see that we are living in the age of the exponential, something for which we have little intuition. At our current rate of increase in consumption, even if our proven reserves of oil resources were five times what they are, they would buy us only an additional thirty to forty years.*

So we have to work hard and fast to make the transition from the use of this vertical energy and move to the use of horizontal energy, the sunlight that falls on the surface of the earth.

We cannot trust nuclear power plants when the fossil fuels are gone, not only because they are uneconomical. We can't trust nuclear power, because it poses a social problem as well as a technical problem. It is certain to bring about the loss of many of our cherished freedoms. How, for example, are we to keep the spent fuel of a reactor, the plutonium, from falling into the hands of terrorists who might make a nuclear bomb? It has been demonstrated to the satis-

*See, for example R. A. Kerr, "How Fast Is Oil Running Out?" *Science* 226, (1984): 426. Also, D. Meadows, D. Meadows Randers, and Behrens III, *The Limits to Growth*, New York: Signet New American Library, 1972.

faction of several nuclear physicists that a smart student in nuclear physics could construct a concealable nuclear weapon capable of wiping out New York City or all of central Kansas.

We have talked of peaceful atoms. In a certain limited sense there are peaceful uses for radioactivity, in hospitals and in research. But the larger truth is that a-tomic means indivisible. To believe that we can divide atoms and unleash their power on planet earth in a peaceful way is *hubris* exactly as the ancient Greeks understood it: the introduction of a pattern into the world that disrupts other patterns that are not of our making, but that we are nevertheless dependent upon.

Clearly, time is now our most precious resource. Remember Marshall Stanton's words: "Nowadays Christians are to faithfully manage their financial, *temporal*, and personal talents and treasures" (italics mine). Including the temporal, or time, was a wise and inspired decision.

To build a sustainable society, we have to rid the world of these weapons. We have to bring the world's population under control. And we have to develop sustainable ways of producing food, clothing, and shelter. All of this begins with the land, with the garden, with people in community. Agriculture must become a primary concern, whether we are farmers or not. In a very real sense, we are all farmers, all of us who eat.

To have a garden, we must have soil. And the soil has to be protected, not just because it is part of the commandment to do so but because it is necessary to life on the earth. The soil must be kept where it was placed, not allowed to run prematurely to a graveyard in the sea. It must be kept free of chemicals that did not evolve with our tissues.

Intimately connected with the need to save our soils is the need to conserve water. Each year Kansans are currently using twice as much water from finite reserves as they use from the water cycle. In other words, we are mining water. Water from the Ogallala aquifer, west of Salina, Kansas, and extending from South Dakota to Texas, is being withdrawn at a rate far beyond replacement levels. In fact, a quantity is being withdrawn each year that equals about forty percent of the total annual volume of the Missouri River at Kansas City. If we were to pull the same quantity of water out of the Mighty Mo during the three and one-half month growing season, the Missouri would be dry below the suction pipe.

We are engaged in this short-run production partly to offset a terrible and growing balance of payments deficit due to the importation of oil. In other words, we mine water to grow corn in order to buy mined oil. That is not sustainability. We grow that corn in part to fatten beef. One pound of beef requires eight thousand pounds of Ogallala aquifer water, accumulated over millennia. In Iowa, five to six bushels of soil by weight are eroded away for each bushel of corn produced. This erosion is the product of an economic system that either regards nature's storehouse as infinite or else simply discounts the future.

It seems the promoters and apologists for this economic system have never heard of justice between generations. It is time that we seriously question our economic system, much of which is, after all, based on greed and envy. Now, lest you think I am unpatriotic, consider that there is a big difference between the economic system called capitalism and the political system called democracy. Because I believe in democracy, I have come to regard capitalism as un-American. Capitalism detracts from our democratic ideals partly because it destroys free enterprise. (By definition, capitalism depends on economic growth that must come from exploiting earth's resources and from forcing more and more people to provide services.) With finite resources, the accumulation of capital means that resources fall into the hands of fewer and fewer persons, and the freedom to be enterprising becomes restricted to those few.

The Soviet socialist system is no better, with its form of state capitalism. In the United States, individuals and corporations rip off the environment. In the Soviet Union, the people, as a group, rip off the environment. The result is the same. We in the West are fortunate in that we still have the chance for free and open discussion.

As we begin to move our technology and life-styles to a solar future, we have to begin to think of new economic arrangements. An area in which I am working and thinking has to do with using a natural ecosystem as a model. Consider nature's local economy, the native prairie. Here is a working model of the way materials and energy flow in order to support a variety of living creatures. It is neither capitalistic nor socialistic, but has been at work on planet earth for at least a billion years. It has been tried and tested since early life forms appeared.

Some of the elements of the Judeo-Christian message lie in this economy of nature. The model is inherently biological. The two major models at work today, capitalism and the Soviet and Chinese brands of socialism, are *industrial*. Christ's metaphors are biological or cultural. He spoke of the vine and the branches, of fish and fishermen. His parables have to do with what is alive, not dead wood or iron or bronze. The Christian message, like an ecosystem, is about process. In an ecological sense, the cross symbolizes a willingness to die so that the continuation of life might be served. Now we must extend our love to the unborn if we are to serve eternal life.

The establishment of a new economic order will require nothing less than the full citizenship that our founding fathers expected of us, which includes the obligation to speak and to participate in communities, in neighborhoods. To build a sustainable society will require nothing less than speaking our minds in wholesome, creative, and responsible ways, moving power from Washington and Topeka back closer to the land, to communities. I think it requires an explicit declaration, at least to ourselves, that we have joined the fight and that our lives' work is laid out before us in a fundamentally different way than if we had not joined the fight. That does not mean that we become, necessarily, full-time specialists in environmental and peace activities. We can be farmers, doctors, lawyers, or teachers. But rather than be known by our careers, we would arrange our lives in such a way that we would work to make the transition to a solar-oriented and peaceful future possible. The role of citizenship needs to be given greater emphasis than individual careers or professions.

We all need to spend less time watching television and more time looking at the geography of natural resources, asking questions about who owns them and how much of each there is and how fast they are being used. We have to become better students of the arms race, of federal spending, of what is happening to the poor here and elsewhere. We should find out for ourselves what is going on in Nicaragua and ask how we would feel if the Nicaraguans mined New York harbor because they did not like our foreign policies.

We hope that our daily work at The Land Institute contributes to this end. We hope that our plant breeding and ecological experiments contribute. We believe that to develop perennial grain crops so that soil won't have to be disturbed every year and subjected to the forces of wind and rain is a patriotic act. So is our work with plants that produce natural herbicides and that fix their own nitrogen—plants that require less fossil fuel input. Our wind machines and solar collectors represent our commitment to develop renewable sources of energy.

Once we join the fight to save the planet, we won't fall for President Reagan's question of whether we are better off now than four years ago. We recognize that our progress as a species does not have to be defined in terms of wealth or material and physical growth any more than our progress as individuals has to be defined in terms of physical growth. Physical growth of the body reaches a limit, but the character and soul of the individual continues to grow, or at least has a chance to continue, often to our last breath. It is simple minded to define our well-being in material terms, when that well-being has an aesthetic dimension, an intellectual dimension, a moral dimension.

If we take seriously the pronouncement that we are "stewards of the mysteries of God," we have no choice but to join the fight to save the planet. That means, in part, that the eyes-to-acres ratio must improve and that people in community must acknowledge that the highest calling of all is to stay at home and work daily to meet the expectations of the land.

Hell Is Now
Technologically Feasible

These last ten years of the nuclear age should convince even the die-hard atheist that the essence of the two-thousand-year-old Christian message has not been altered. Christians have always been directed to love their enemies, to pray for those who use them spitefully, to love their neighbors as they love themselves, to work in behalf of the Prince of Peace. If not, the warning goes, there will be hell to pay, a heat death, no less.

These old commandments are hard to follow; even the Christian world has done a poor job of it. In modern times, we have even come to believe that there is no point in following these commandments, for when we die, it's simply curtains for us. Since our bodies first took on meaning when we came into the world, why shouldn't that meaning disappear with death? Since we return to nothingness, there is no penalty for bad behavior. Many people throughout Christendom have come to regard the Christian code of conduct as something like an annuity. The benefits terminate at death.

Suddenly, in the last two-and-a-half percent of the Christian era—that's the last half of *this* century—things have changed. Suddenly, it has become painfully clear that all the requirements of the ancient commandments must be more rigorously followed than ever before if humankind is to survive. It is almost as though God has said, "I threatened them with heat death if they didn't follow the instructions in the old program, but that threat hasn't worked. Because they have not loved their enemies and their neighbors, they must now risk the threat of the 'second death,' extinction. Since they don't care about themselves, we'll see if they care about the species, or for that matter about all life on earth."

As David Ehrenfeld has suggested, it is as though the Creator has designed a trap so elegant that none will escape; not the president, not the generals, not the rich, not the ordinary businessman or factory worker, not the farmer or the Quakers. Either we get with the specifics of the ancient commandments, beginning here at home, or the "second death," the heat death from the ancient fires of the universe, is inevitable. We are *required* to love our enemies, not stockpile weapons against them or point missiles at them from land or sea or air or satellite. Though the technology for a widespread heat death is less than a half-century old, the warning is nearly two millennia old.

Jonathan Schell, who first used the expression "second death" in his book *The Fate of the Earth*, makes an airtight argument for disarmament, beginning with nuclear weapons. The careful thinkers on this subject eventually conclude that in the nuclear age, we must remove *all* weapons if we are to avoid another arms escalation. The energy to do so can only come from loving our enemies and our neighbors. We must love our enemies both at home and abroad. We must love our neighbors. The threat that backs these Christian commandments is absolute. Of course, it is difficult to back away from the threats of the Soviet Union, but ten thousand years of existence under the worst imaginable authority—a Hitler, or a Stalin, or any other political or social absurdity we might be forced to cope with—doesn't compare with extinction, or even life in a nuclear wasteland. Human will is powerful and its devices sufficiently complicated to break any political stranglehold, but human will would be powerless on a densely radiated earth.

It won't be easy for me to love my enemies or even all my neighbors. I have been angry with most politicians, mad at the generals—the world's leading terrorists—for twenty years now, angry with my

colleagues in the universities, where potentialities rot and few seem
to care. Greed and envy are unbecoming in their own right. What is
one to do when it becomes unequivocally clear that these two vices
are primarily responsible for reducing options for the unborn? And
yet, we are required to love *those* so possessed. We're in for a long
pull.

I have wanted to underscore what the economist Paul Hawken
says: that the economy that everyone says isn't working, *is* working—to the advantage of the rich and to the disadvantage of the rest.
Because billions of petrodollars no longer circulate in the U.S. economy, because we have become resource-poor relative to our level of
consumption, the economy now sends signals we choose to misin-
terpret. What is happening now in the economy is the consequence
of individual action. This is not an immutable force moving, like
some law of physics, across the landscape. When three-hundred—
or even thirty—year-old trees are cut for lumber or chips from a dry
California mountain, the economy pays no attention to the tree's age
or to the fact that an entire generation of the descendents of loggers
will have to go somewhere else to make their livings. But one day
the chickens will come home to roost. What is happening now in the
economy is a small penalty for the past and a faint foreshadowing of
the future. I know that I am responsible for fueling this economy too.
But it is one thing to be caught in the system—for there is no life outside it—and quite another to defend it. I find it hard to love those who
defend it, and I confess that because I have not loved all my enemies
and all my countrymen and neighbors I have increased the likelihood
of a new eternal hell on earth caused by nuclear weapons. I believe
this in my heart of hearts now, for as a biologist, I believe that everything is connected to everything else, that Garrett Hardin is right that
we can never do *only* one thing. Without knowing the probability, I
believe that every act either increases the chances of nuclear war and
the likelihood of extinction or decreases it. An elegant trap has been
set, and the ancient program abides.

How to Avoid
Building Pyramids

The human race was born out of nature and it is out of nature that the human race and all life is sustained every second of every minute of every hour. This has to be our beginning point in thinking about the different economic orders that now exist on the earth. It turns out that neither of the two dominating economic ideologies, neither capitalism nor Marxism, is fit for our planet. Fundamental to both of these nearly identical twin orthodoxies is a barbaric plunder of nature. Marx especially was explicit about this. He saw this plunder as necessary and inevitable if progress is to be made toward a better world for humanity. He endorsed this plunder wholeheartedly, perhaps because he was a city boy who believed in "historical necessity."

He also believed that economic laws were as immutable as physical laws. (Perhaps the laws are immutable, but the assumptions or variables we plug into the equations are not.) He was a devotee of the idea of centralist planning, with little concern for the consequences to the periphery. It wasn't just nature that had to be sacrificed in the name of human progress. Native cultures were to be sacrificed too, for they were mere anachronisms. Historical necessity meant that the destruction of these cultures was inevitable during the creation of the material basis for civilization. For example, he commented once that England had a double mission in India: Destroy the old

Asiatic society, and in turn lay the material foundations of Western society. (To his credit, Marx did recognize the finitude and value of the soil. He wrote that "all progress in capitalistic agriculture is a progress in the art, not only of robbing the laborer, but of robbing the soil.")

Couple Marx's materialistic view of nature with the notion of John Maynard Keynes, the architect of modern capitalism: "Foul is useful and fair is not," said Keynes. "We must have foul a little longer." He meant that we must use greed and envy in order to open up the mines and well heads and get the raw materials for consumption spread around the earth. When there is enough for everyone then we can suspend greed and envy.

To suggest that someone of Keynes's intellect and stature was being naive or silly somehow seems inappropriate, but this idea of using long-recognized vices of humans to plunder the planet until some time when we will suspend these vices at will seems just that: naive and silly. It is also dangerous, for giving a green light to exercise traditional vices will mean that many will suffer in the long run for the benefit of a few in the short run.

Marx has had his impact, for we have come to act as though economic laws are as immutable as physical laws. The problem is, we often confuse our assumptions with the laws. Even environmentalists who should know better think they are using economic arguments to support their case. No economic system exists by itself. Rather, any economic system is the consequence of the collective values of the culture. Change the values and you change the system. Take one extreme example. Most humans desire gold; therefore gold has value. If no one wanted gold, it would be worthless. More importantly, because humans have a penchant for gold, cultures have worked out many different ways to allocate that gold to meet some desired end. Though the Aztecs shaped gold into beautiful coins and figures of adornment, the conquistadors cared only about the marginal value of the gold, so they melted down many of these works of art. They cared not a whit for the add-on value assigned by the native cultures.

I am making such a fuss about all this because to call the assumptions "laws" and then to treat them as immutable as the laws of physics does us a disservice. On the one hand, it makes us feel that changing human values is pointless; on the other, it makes us overly optimistic in hoping that the market might destroy the possibilities of building nuclear power plants. This religious faith in economic determinism keeps us from discussing the ultimate consequences of centralized planning and power delivery. Just as important, it keeps us from confronting and enlarging our own value. Hundreds of Kansans knew that economics did not support the Wolf Creek nuclear reactor that recently went on line in Kansas. It is a matter of record that we foresaw nearly every step of the scenario years before it was played out. We predicted the horrendous cost overruns during construction. We predicted that the utilities would want to pass on the cost of their stupidity in planning to the customers rather than to their stockholders. We feared that the Kansas Corporation Commission would probably cave in toward the end and bail out the utilities that had overestimated how much power was necessary for the future. Watching the inevitability of it all was like watching a dramatic tragedy unfold on the stage. It was devastating for all of us who opposed the monster, but many important things happened during the fight. We formed an alliance with protestors in other states. We learned even more about the diseconomies of nuclear power and became radicalized about the nature of our economic system. We realized that humans in high positions were no smarter or wiser than we. We learned that we have to work harder to elect better representatives. We learned what it means to have a politically ambitious governor. We learned that we have to have more ecologically aware corporate commissioners.

Some people say that nuclear power will die under its own weight, that the market will not support the building of nuclear power plants. For the moment, this is true, in this time of abundant coal and oil and natural gas, and a smattering of solar installations. In this interim, our challenge is to begin to teach, experiment, and think hard about running this culture on sunlight, for if we fail in this task, then any number of cultural arrangements and numerous forms of political suppression will be put in place by those who support the idea of "historical necessity." According to Herodotus, the pharaohs enslaved and persuaded tens of thousands of people to build pyramids to house dead kings. In the twentieth century, we "objective" observers might regard pyramids as stupid projects and wonder how so many people could be organized to pile so many stones in one place. Apparently numerous off-season Egyptian farmers would join the labor force for pyramid construction, partly for religious reasons. Several of the people, at least, believed that their immortality was hitched somehow to the immortality of the pharaohs.

Suppose that thousands of slaves, though they may have been "unbelievers," acknowledged the economic necessity of their work: if they wanted to eat, they had to keep piling on the rocks. While this may have been an acceptable religious goal, it was certainly a terrible social goal. Those pyramids are the result, in large measure, of the vision and will of a few elite kings and priests.

Where was the market in the pharaohs' time? Exactly where it is today—serving "necessity." Whether an economy serves the necessity of a religious hierarchy or the necessity of the stockholders in Kansas Gas and Electric is an incidental detail. In neither case is it serving the necessity of the people.

And so where is the culprit and how do we begin to manage our exodus from the somewhat benign tyranny of modern-day primordial pharaohs? I think that we begin by repudiating the capitalistic and Marxist notions of historical necessity. The pharaohs built pyramids and the utilities nuclear power plants. Both were and are devotees of centralized power generation and delivery. Whether this power is political power or electric power is unimportant for the two are so closely tied. Ultimately then, the battle is between the centralists and the peripheralists.

By centralists I mean those who favor centralized control, centralized planning and centralized dispersement of services. The peripheralists, or decentralists, prefer greater control over their local economies and greater dependence on community loyalty. They are what E. F. Schumacher calls "homecomers." They want to be connected to the land, not just owners of it or extractors from it. Their resilience comes from a sufficiency of people. If we are serious in our pursuit of a sun-powered culture, then we must spell out the differences between the loyalties and affections of the centralists and the peripheralists.

We have to begin with a basic reality: we're talking about a sun-powered culture. Sunlight falls in a dispersed manner over the surface of the earth. There is an optimum radius which defines the circle over which sunlight can be organized for human use for various purposes. The area of that circle and the number of people in it depends on local conditions of soil, rainfall, temperature and humidity. That optimum radius or area and the optimum number of people in each place can define our local politics as it has defined the politics of native cultures in the past, and as it defined the establishment of rural culture following settlement in America. The model for this new economic order has not been worked out but it is now in the making. The people at work on this new order will generally mistrust the centralized planning that homogenizes both landscapes and people. They will acknowledge that the human race was born out of nature and that it and all life *are* sustained by nature every second of every minute of every hour. They will acknowledge that it is a terrible and cruel lie that we must plunder nature in the name of historical necessity or to satisfy the demands of the market. They will acknowledge that native cultures and rural communities are not anachronisms and that their destruction need not be inevitable. They will understand that because these cultures are peripheral they have likely accommodated themselves to living locally and thus constitute a source of information for how to live in a dispersed manner on a sun-powered planet. Finally, they will see that to use the vices of greed and envy in order to achieve some short-term economic benefit is a kind of immorality. To institutionalize these vices amounts to an immorality of culture that will not be easily abandoned. We must call these vices what they are, refusing to allow them to be used as levers for some "ultimate good" as a matter of "historical necessity."

Pre-Copernican Minds
of the Space Age

Long after Copernicus straightened us out on the relationship of the earth to the heavenly bodies, we continued to think of earth as "below" and of "heaven" as above. Until recently, most of us regarded the heavens as out of reach, at least for as long as we were alive. Following the famous failure to reach heaven by means of the first massive public works project, the Tower of Babel, we gave up on trying to reach the heavens by physical means until the so-called space age. The Babel project was abandoned, the story goes, because God came down, saw what the people were up to, and realized how "nothing will be kept from them that they have imagined" (Genesis 11:6). He took quick action. He confounded the language of the workers and dispersed them; no great task, really, for they must already have been arguing about the location of God's residence. And if God resides in heaven, he must live at a certain altitude. At this point, the workers would have been likely to disagree, for the eventual height of a structure dictates the width of its base, the depth of its footings, and so on. The workers must have argued among themselves on the specifics of the project.

The larger point is that in pre-Copernican times, the notion of an earth below being watched over by God above was widely held, and understandably so. For these believers, the height of heaven was a matter of individual opinion. Now, in the space age, our impulse to get to the heavens is still powerful, and we are employing our most sophisticated technology in our efforts to get there. NASA has spent billions, testimony to the possibility that this impulse in us modern folk is even stronger than it was in the builders of the Tower of Babel.

I suspect that we have such a penchant for space travel because we are but third and fourth generation descendents of wilderness tamers. The space program is our modern wilderness to be conquered, our new frontier to be colonized. We have set out to colonize space in the same manner that we have sought to colonize the cell, DNA molecules, and even the inner recesses of the atom. We are a middle-ground species that colonizes in both directions.

Colonizers always live under powerful illusions and an inescapable ignorance. It is a peculiar sort of ignorance that causes us to forget that we have always lived in the space age. The earth is already in space, after all, and some of the speeds at which the earth already moves through space are likely to be unattainable with our technology. The earth's rotation is not so fast—around one thousand miles per hour at the equator—but the speed of our orbit around our star is about 66,600 miles an hour. From here on, the speeds become truly unimaginable, for our entire solar system moves among the neighboring stars at a clip of over four hundred thousand miles an hour. The entire set of stars moves in the galaxy of the Milky Way at more than half a million miles per hour, making one revolution every 200 million years. Given the speed of our modern rockets, it is important to ask what are our chances of discovering another planet somewhere that would have precisely the right mix of gases and gravity to accommodate our needs. If we can't live on another planet, except for brief periods, then our space program can have no value except as a mine to service an extractive economy.

We do live in the space age, but in the same manner as did Jefferson and Charlemagne, Jesus and Aristotle. We are in heaven *now*. As Elizabeth Barrett Browning wrote:

> Earth's crammed with heaven,
> And every common bush is afire with God;
> But only he who sees takes off his shoes.

Ridding our minds of what we believed before Copernicus is long overdue. And maybe the real work of protecting and maintaining our beautiful craft cannot begin until, like Moses before the burning bush, we are moved to take off our shoes. Shoes on, we can spend a lot of money to go nowhere slowly with our modern rockets; or shoes off, we can stay at home and go nowhere fast—just as our ancestors did.

Land Wisdom
vs. Lab Success

"Aggregation of power" is really my subject here, for what I am about to talk about is an emerging problem for agriculture as the result of such aggregation. Much of it is within the university, some within the U.S. Department of Agriculture (USDA), and some within private industry. The story really begins in the mid-1940s but I am going to jump forward a decade.

In the mid-1950s, biology departments were mostly in small colleges. Unlike the universities and major colleges, most of these small college programs could not afford the luxury of separate departments of botany, zoology, entomology, and bacteriology. This was in the good old days of biology. Of course, schools with their separate departments *did* have problems. It was not unusual, for example, for botanists to refuse to speak to zoologists, and the other way around. Such fragmentation led students to believe that the differences between plants and animals were more fundamental than they are. There were some good things in this division—for example, it kept botany from being swamped by a pre-med program—but this separation came to a quick end as institution after institution placed all the disciplines under biological sciences or within a department of biology. Some universities did not collapse their departments under one empire, but most did.

This aggregation of power can be conveniently traced to 1957, the year the Russians threw the first Sputnik into the sky and sent a seismic wave through all of American science. We suddenly felt old-fashioned and in need of catching up with the Russians, so we spent more money on science education and research. Science programs expanded and the era of big science got a boost, literally, from a Russian rocket. This was the beginning of the end for numerous small subdivisions in biology. In a way, money did them in. Scientists began to specialize and to establish little empires. To be honest, more than money was involved. Numerous professors promoted the *biological* argument that the difference between prokaryotes and eukaryotes was more profound than the difference between plants and animals. In other words, blue-green algae and bacteria were more different from redwoods than redwoods are from lions. The old distinctions between plants and animals were said to be too arbitrary.

Evidence coming in from other directions as well began to make this division embarrassing to biologists. Animal ecologists were aware of the dependence of the animals they studied on the surrounding vegetation. For both animal and plant ecologists to be under the same umbrella in the organizational structure of the university seemed to make sense.

So in the early 1960s, the cracks began to widen in the rigid organizational structure of the traditional disciplines in biology, cracks that the academic deans worked to their advantage in academic empire building. They were *doubtlessly* tired of the squabbling among the various heads of botany, zoology, entomology, and bacteriology. By putting them all under one organization, the squabbles over budgets could be settled below the dean's office. Had there not been a charismatic Kennedy, a civil-rights movement, a Vietnam war, a proliferating counterculture, the public might have learned of this not-so-quiet revolution in the structure of biological learning and research. It may have made little difference that the public didn't know, but what was to follow was to have profound implications for those of us interested in sustainable agriculture.

And now we must consider another historical theme of equal importance, a theme that would eventually intertwine with the one that I have just discussed and have a tremendous impact on biology—as, I am afraid, it will have on agriculture. We may say that this second theme began to develop in 1944, the year that Avery, MacLeod, and McCarty published the results of their experiments that suggested that DNA and not protein was the chemical responsible for heredity. Less than ten years later, in 1953, James Watson and Francis Crick reported that the structure of the DNA crystal was a double helix. They got the Nobel Prize they were after and the terms DNA and double helix entered into common usage.

The momentum for molecular biology, as a field, was now well underway. Many of these new biologists came out of chemistry. They did not come out of the tradition of biology. Most were not steeped in the biological lore, nor, I suspect, did most of them care to be. Avery, MacLeod, and McCarty, Watson and Crick came to be names that crowded out other names and other concepts in undergraduate biology courses, as professors upgraded their notes and changed to new texts. Even though the budgets for scientific research and teaching were expanding during this post-Sputnik era, even though new buildings were being built to accommodate this growing scientific establishment of professionals and equipment, some of the traditional schools of thought suffered.

Countless botanists were bitter because organismic botany, in particular, suffered. It was regarded as too descriptive—not analytical enough. As many of the old plant ecologists retired, they were not replaced, and their labs were redesigned to accommodate the new, young breed of molecular biologists. After all, as one old scientist said, there was a "killing to be made on DNA."

I know what this fever was like. As a graduate student in the 1960s, taking a course in biochemical genetics, I anticipated with fellow students the most recent issue of the *Proceedings of the National Academy of Science* in order to learn what discoveries had been announced in the past month. We waited with fascination to learn how this language of life was arranged. I have to admit that it was a most exciting time in the history of biology.

During this fever, budgets for scientific hardware went sky-high. Electron microscopes, which can now cost half a million dollars, became absolutely essential, as did fast and accurate weighing equipment and growth chambers controlled by computers. Department heads and deans had become responsible for multimillion-dollar empires. These managerial academics had the best of both worlds. Not only could they associate with the world of high intellects, they could measure their worth in the prestige and importance that adhere to large budgets. They could compare themselves to the middle, if not the top, executives in major corporations.

For a quick study of this movement to expensive high technology in science, check the ads in the weekly journal *Science* over the last twenty-five years, and note the increase in technical sophistication and expense. Look, also, at the ads for positions available. "Cutting edge" science became very expensive. We have known that it is costly to the practitioners and administrators, for such science becomes the fuel for hubris, but few could have seen what the revolution in molecular biology would mean to botany. Hindsight is always 20/20, but shouldn't we have wondered where all those two- and three-year post-docs who were doping out the code would go once their post-doc was over? Of course the momentum was there to accommodate *them*. They had prestigious bibliographies. They had worked in the labs of Nobel laureates and near-laureates. They found good jobs in major universities. And since it is a monkey-see-monkey-do world, they, in turn, took the best graduate students available and put them to work on even more "cutting edge" science. But almost without notice, the era of discovery moved smoothly into the era of manipulation, until suddenly we had new household words and phrases such as *gene splicing*, *gene stitching*, and *DNA surgery*. We were told early on and we are being told now that this new biology will help us cure cancer; and that it will make nitrogen fixers of temperate cereal crops when we stitch in legume genes to accommodate the nodule-forming nitrogen-fixing bacteria.

I have just described the revolution that went on in biology. What I have not yet described is the illusion of a revolution or a pseudo-revolution that went on in agricultural research. During this great change in biology, there was a tremendous change in the structure of agriculture worldwide, mostly because of yield increase. A so-called green revolution was taking place in Mexico, in India, in Asia. The casual observer might conclude that the revolution in biology was responsible for the revolution in agriculture, that they were proceeding in lockstep to increase the food supply for a hungry world.

The two revolutions were, however, more or less independent of one another. Yield increases were the result of the widespread use of commercial fertilizer and irrigation and the increasing use of pesticides. Crops were designed to be less discriminating in fertilizer uptake—in other words, the genetic mechanisms responsible for the orderly uptake of fertilizer were destroyed. Instead of worrying about this loss of genetic information, we said that such plants had "high fertilizer response." What is important to appreciate is that at this point molecular biology had little or no impact on agriculture. Because more researchers had understood the statistical models of Sir Ronald Fisher and others, increased sophistication in experimental design led to a more efficient means of selection and an increase in the number of inbred lines and marker genes. All in all, it was the establishment of the international research centers, the growing sophistication of some of the major seedhouses, and the interaction of those companies with the geneticists and agronomists in the land grant universities that brought on this so-called green revolution. Was it ever dramatic! In fifty years, the yield of corn went from a national average of thirty bushels per acre to one hundred. But it was not an era of discovery so much as an era of implementation.

Eventually, we came to the era in agriculture, in the mid to late 1970s in which an increase in fertilizer was not proportionally met by an increase in yield. The fertilizer curve line was going up and the yield line was flattening. Moreover, farm yields were coming closer to the yields in experimental trials. In other words, research results were no longer far ahead of field results.

It was inevitable that these converging lines would be noticed. Agricultural researchers would coast for a while on their past dramatic achievements, but a few impatient scientists, mostly outside of agriculture, would point out that the technology that breeders had employed to bring about the record yields was about milked dry and that we should start now to implement the new science and technology that had been established in biology. This would give us a new knowledge base to exploit for the purpose of feeding an increasingly hungry world, and it would shore up our ability to produce for a future export market.

In June of 1982, there was a landmark meeting—a potentially dangerous one, I think. Dennis Prager, a physicist who started work as a policy analyst at the Office of Science and Technology Policy during the Carter administration, met with the Rockefeller Institute's John Pino. They and a few other people held a conference that summer at Winrock, Arkansas, and concluded that the land grant institutions were lagging in their basic research and were therefore desperately in need of the new knowledge in biology. The best work, they concluded, was being done outside the agricultural system. As a result, there is now a move to upgrade agricultural research.

In response to this high-level conference, the wheels of bureaucracy squeaked forward. Orville Bentley, a biochemical nutritionist and former dean of the University of Illinois College of Agriculture, was the assistant secretary for science and education in the USDA. He presided over the Agricultural Research Service, the Cooperative State Research Service, and the Extension Service. He was quoted in *Chemical & Engineering News* (November 22, 1982) as saying there is a change taking place that is more rapid than gradual. He said that "there will be a swing toward mobilizing our resources toward biotechnology, genetic manipulation. Two other important areas are resource utilization and protection of soil and water." But he goes on to say, "I still think the driving force will be efforts to increase yields, productivity and production as a way to keep the level of technology high." Bentley does say that there will be investigations into the feasibility of diverting some of the farm subsidy money, $7 billion or so, toward research and conservation, but admits that this will involve a fight.

William E. Marshall, a biochemist who is technical director in the development sector of General Foods Corporation, has been studying federally funded agricultural research as a member of President Reagan's Task Force on Cost Control. He contends that the best work is being done outside the agricultural research system, and he concludes that "what's needed is to bridge the gap between molecular biology and the future of agriculture."

During the past twenty years, the molecular biologists who were taking their post-docs during the 1960s have professionally cloned themselves. The modern-day descendants of the new breed in the sixties, like their predecessors, may never have had a field biology course, never milked a cow or goat, maybe never driven a tractor. But they are looking for work. There are only so many pharmaceutical houses, only so much interferon to be made, only so many who can work at tricking bacteria to make insulin. There they are, credentialed, knowledgeable of the equipment, toned up on the literature, ready to roll.

What they have in mind is currently limited, but the future is boundless. They plan to turn grasses, for example, into plants that will fix nitrogen as readily as some of the major legume crops. They hope to introduce genes for resistance to various insects and pathogens. They hope to boost yields. Sounds good, doesn't it? Their agenda for agriculture is difficult to argue with. They are presenting a world of the future that makes one think that one is watching the upbeat ads during the Superbowl. But it is clear that what they are doing is trying to write large the last fifty years of agriculture. They are offering the "specific problem–specific solution" approach as the infallible recipe. This approach assumes that everything outside the specific problem for which they intend to splice in a solution can be held still, that nothing else will wobble; or if it does, that they can splice in a correction for that, too.

All of this is high tech research and I suspect that any outfit that gives you a crop with a spliced-in gene is going to demand a patent and some kind of a royalty payment. It is doubtful that their primary concern will be the high energy cost of American agriculture. One also doubts that they will care greatly about the national and global soil loss problem.

Of course, agriculture needs very little of what molecular biologists have to offer. But none of us wants to be merely against something; we want to be for something as well. We are fortunate, then, that another kind of change which offers some possibilities has been going on in biology, a change, though so far scarcely noticed, that can help agriculture, though it will not do so automatically.

This change comes from a synthesis of several fields, from people who have had various motivations. They are taxonomists, ecologists, and geneticists, and they have been putting together a new field that might be called, for want of a better name, population biology. The primary contributors to this field have been plant ecologists and population geneticists, people with interests in evolutionary biology. They study the strategies that species employ to survive and multiply on the land. They study relationships in the production and allocation of energy in plants: does a plant send the harvested sunlight to the seed or to the root to overwinter? They study senescence in plants, the mechanisms of interaction among plant species, the diversity and natural dynamics of populations. They are interested in weeds as colonizing species, insect interactions, and the role of pathogens.

Especially in the U.S., these researchers have been just as interested as the cutting edge scientists in molecular biology in accumulating knowledge for its own sake. One difference, of course, is that the pharmaceutical houses have little use for population biologists, who can't make insulin or interferon. None of these professors will be invited to join a new company like Genentech with a starting salary of $100,000 per year and all sorts of stock options. Some of the motives of these population biologists and ecologists may be the same as those of the molecular biologists. Too many have used their degrees and their bibliographies as passports to privilege rather than responsibility. Nevertheless, they are more in the tradition of the long distance runners in research than they are like the sprinters in molecular biology. Their roots are in traditional biology. They can trace their academic ancestry back to many of the old names. They know the lore of their heritage in the long tradition of biology, the kinds of stories you pick up around a department, up and down the halls. They know the work of Asa Gray, Bessey, Weaver, Clements, Stebbins, Dobzhansky. Darwin and Liberty Hyde Bailey are part of their being.

The work of these plant population biologists, evolutionists, or ecologists is admittedly still at the "knowledge for its own sake" level. But they are working so high above the level of the individual gene that the nature of their research is fundamentally different from that of the molecular biologist. What they have accumulated and what they have to offer is what those of us interested in a sustainable agriculture need to pay attention to, for the sustainable agriculturist begins with the notion that agriculture cannot be understood in its own terms—that it comes out of nature. The test for this is the question whether a crop plant should be regarded more as the property of the human or as a relative of wild things. If it is viewed primarily as the property of the human, then it is almost wide open for the kind of manipulation molecular biologists are good at. If, on the other hand, it is viewed as a product of nature primarily, as a relative of wild things, then we acknowledge that most of its evolution occurred in an ecological context, in nature, the design of which was not of our making. I want to underscore the fact that the population biologists are at the other end of the spectrum from the molecular biologists. They may admit that humans learn faster than nature but they acknowledge also that nature is hard to beat because she has been accumulating information longer. Most of the mistakes of nature have been corrected over time.

Agricultural research can benefit more from this inherently broad tradition than from the narrow innovations of the gene splicers. A new agriculture must come from people who are students of nature at the ecological level. For after all, a natural ecosystem, like a prairie, sponsors its own fertility, recycles its nutrients, avoids the epidemic from both insects and pathogens, and does not lose soil beyond replacement levels. Many of the people who make it their lifetime occupation to study the kinds of ecosystems that feature all these elements of sustainability will have to be moved from the pursuit of "pure" knowledge. We must ask some of them to turn their heads and hearts to work with us in the development of a sustainable agriculture.

Our first task is to say *no* to the imperative of molecular biology. We must not allow it to horn in on agricultural research, aiming only to increase the size of the production-oriented Leviathan. Second, we must encourage and aid the resistance of the current researchers in agriculture who resent the intrusion of molecular biologists. Third, we must work on the other breed of "pristine" scientists in the biology departments—the physiological ecologists, the population geneticists, population biologists, evolutionary biologists, biosystematists, and the rest of the biologists—and let them know that we think it is time for them to begin the difficult task of working in the area of ecosystem agriculture. Nature can work to our advantage in agriculture. Some work of this kind is already going on. We all know about the work of the Rodale Research Farm in Pennsylvania, the work of Richard Merrill of the Portola Institute in California, that of John Jeavons at Willitts, California, the permaculture work of Bill Mollison, Masanoba Fukuoka's work in Japan, and Sir Albert Howard's work in India early this century. Steve Gliessman is now working at Santa Cruz, Miguel Altieri at Berkeley, and we are at work here at The Land Institute in Kansas.

But all of us have just begun. We need to take the ecological knowledge that has been accumulating for the last thirty years or so, and think on its applicability to agriculture. It is going to take a concentrated effort to move this knowledge into the Agricultural Research Service of the USDA, into the land grant institutions, and, most importantly, onto the farm.

What those of us interested in sustainable agriculture need always to keep before us are the questions: How are we going to run agriculture and culture on sunlight? What are we going to do when the oil is gone? What are we going to do to stop soil erosion? Ecosystem agriculture has answers to all of these questions. Molecular biology has few or none.

It might be argued that the fields of population biology and ecology are so complex and that so little is known of living things and the physical-chemical world that surrounds them that to concentrate at the ecosystem level, rather than at the population or organism level, will be impractical. Work at the ecosystem level could be rejected even if our goal is to save soils, prevent chemical contamination of the countryside, and get farms to sponsor their own fertility and energy. The argument could be made instead that the payoff from the likes of gene splicing is sure and promising for solving the immediate problems of agriculture, and that molecular biology is mature enough, that we should get cracking now.

I don't think so. The synthetic fields of population biology and ecology are just as mature. They haven't received the same amount of media coverage. They haven't been featured in *Time* and *Newsweek*. But consider the barriers molecular biology must overcome to be able to deliver on the promises that have to do with production only, promises that never include the notion of sustainability.

As I mentioned above (in "Biotechnology and Supply-Side Thinking"), with few exceptions, and they are exceptions because of certain anomalies, the gene splicing work to date has featured the relatively simple prokaryotic organisms, the bacteria and their associated viruses. Such organisms are several orders of magnitude simpler than the kinds of cells nature has used to make redwoods and lions, lilies and people, and (with the help of humans) corn plants and Holsteins. For gene splicing to be useful at this level, there must be a method of incorporating the gene into the entire genetic complement of the recipient species. This will be no small trick, but let us assume that it can be done.

First, the team of molecular biologists must know what gene or genes they want to transfer from one creature to another. Next they must find a source. Then they have to be able to extract the small amount of DNA representing that gene or genes out of the rest of the DNA in a complex cell. So far, maybe so good. What if the transferred gene fails to work in the new environment? They will have to find out why. Most of the requirements necessary for that gene to function in its new and alien world will be unforeseen and unforeseeable. Furthermore, it is unlikely that the newly modified genome (all of the old gene material of the host cell, plus the newly introduced material) can be easily propagated.

So much for the easy part. I call it easy because it involves the most straightforward manipulations imaginable so far. Now for the hard part, the more formidable problems. Because all genes interact to some degree, the traits that are *strongly* influenced by several genes working together will stand as a barrier to the gene splicer. They are still beyond the current state of the art for gene splicing. Professor Dick Richardson, a geneticist at the University of Texas at Austin, points out that some traits such as growth rate are affected by many hormones, including episodal ones that are present for short periods of time in low concentrations. Many of these are only now being discussed. When their existence is known, isolation may begin, but if the genes are from widely divergent organisms, the new host may regulate these hormones in a way that is completely foreign to the implanted gene. For example, the same quantity of a particular hormone produced during development in one creature may yield a very different effect in another.

Professor Richardson reminds us that a gene is often separated into several pieces and located in widely separated places on the chromosome or even on another chromosome. While this is a tricky problem to overcome, it is no more tricky than isolating the various genetic components that regulate a particular gene in question. Once a complete gene and all of its regulators are isolated, there remains the problem of the entire assembly becoming precisely incorporated into the genetic material of the recipient organism. If it isn't incorporated early enough in development and misses being transferred to the offspring, for all *practical* purposes, the splicing has reached a dead end.

Let us assume that all these barriers have been overcome. We are now faced with a problem somewhat similar to what geneticists confronted nearly forty years ago, during the heyday of radiation genetics. This was a time in which numerous geneticists believed we could improve crops and speed up evolution by irradiating the germ plasm and then selecting the desirable products. What that generation of geneticists and plant breeders learned is that they had on their hands the same problem as the previous generation of geneticists who had believed that some biological wonders could be pulled out of the progeny of some very wide crosses. The problem they had was how to get rid of all the variation they suddenly found on their hands, and how to *reoptimize* the desirable traits against such a scrambled genetic background. The background of spliced-in genes may not be so scrambled, but the problem of reoptimization is still there. In other words, even if all the steps are successful up to the point where the spliced gene and its regulators from a distant plant family are transferred, an untold amount of breeding work remains before the genetic background is shaken down enough to accommodate the newly introduced trait and its regulators.

The ecosystem level of biological organization is complex, much more complex than the DNA level of any species, but it is not necessarily more *complicated for the human*. For that matter, the level of the molecule is more complex than the atomic level but molecular biology as a field is not more complicated than physics as a field. At the ecosystem level, if researchers and farmers take advantage of the natural integrities that have evolved over the millions of years, they may be dealing with great complexity. But that complexity may be much less complicated for the human to manage than gene splicing at a much simpler level of biological organization. Ecosystem researchers will simply be dealing with huge chunks or blocks of what works.

If we continue to lose soil, if our soils and groundwater supplies continue to be polluted because of our single-vision focus on production, the day will come when few will care whether molecular biology ever existed as a discipline. There is enough "on the shelf" knowledge now, all in the area of population biology, evolutionary biology, and ecology, to begin to meet the needs of the land and the needs of this species of ours, that was shaped by the land.

Why have we been so slow in getting started on ecological agriculture research? Well, such an ecological agriculture was really not possible until the last ten or fifteen years, until the great synthesis began to emerge, until sufficient knowledge about the workings of natural ecosystems had been discovered.

We still have a great opportunity to do something about the problem of agriculture, but we have little time in which to take advantage of that opportunity. Right now, average agricultural researchers are in their midfifties, about the age of average farmers in America. This means that in the next five to fifteen years, a lot of agricultural researchers are going to retire. Sixty thousand professional slots will open in agriculture next year and there are only fifty-two thousand people trained to fill them, a deficit of eight thousand. There will undoubtedly be deficits in subsequent years as well. What this means is obvious—if we can get people trained in ecological agriculture, we could change the structure of American agriculture very fast, for in another ten to fifteen years, many of these people would move into positions of responsibility. If we fail to produce enough students of ecological agriculture, then molecular biology will win the day.

Numerous problems lie before us, but the day may not be too far off when scientists who have been studying natural ecosystems will begin to talk to farmers as equals, and when farmers and scientists will join together in the common task of learning how to live *decently* on the land surface of the planet and make a *decent* living while doing it. We should acknowledge what molecular biology does have to offer sustainable agriculture. It won't be much. We next have to use whatever energy we can to tilt future agricultural research toward an ecological emphasis. That will be tough because powerful interests are becoming more aligned with exploitative agriculture as each day goes by.

Meeting the Expectations
of the Land

Part of the modern problem in agriculture is that our policymakers, if not the population at large, treat agriculture as an isolated part of the society—a segment in which something has gone wrong. Expensive salvage operations are designed, therefore, around the notion that agriculture is a problem that needs fixing. The phrases that come tumbling out of many of the deeply troubled and the superficial are pretty much the same. We hear statements such as: "Pure and simple, it is strictly an economic problem." "Agriculture is in trouble." "Something needs to be done about the farm problem."

Sure, scarcely three percent of us in the United States are on farms. Farmers are a dispersed minority and have little political clout anymore. Were they a dispersed majority, the farm vote would still make a difference. Were they a concentrated minority, they would be close enough together to hammer out their differences and speak with one voice. But of course they are neither.

While most of the phrases about problems on the farm are true, at least in a limited sense, none suggests that problems on the farm are more the failure of culture than of economics and public policy. Economics can define the problem, but only in part. It won't provide a solution, yet nearly all our public policy decisions are based on economic pressures.

What I hope to offer here is the consideration that some of the problems *in* agriculture are mere derivatives of the problem *of* agriculture, which in turn is part of a systemic problem for the culture at large. This has scarcely been understood in the United States, or for that matter, I suspect, in any other country. We seem to keep hoping for a breakthrough. Note that several of our nation's music stars organized two huge fundraising concerts last year: one for the starving in Ethiopia, and one for the nation's farmers, who have produced too much and so have gone broke.

I see three tiers of problems embedded in the problem of agriculture. And though I want to deal mostly with the middle tier, I'll begin by considering the first. In an absolute sense, the problem of agriculture will probably never be solved. In my view, it is part of the Fall. I suspect that agriculture is at the core of the Fall. We can imagine that the Fall came when the gatherers and hunters *expanded their scale* from patches into fields. This decreased our reliance on nature's wisdom while increasing our dependence on human cleverness. We came to depend more on human knowledge. There are too many people now for us to become gatherers and hunters again and so, after the fossil fuels are gone, we will, with most of the world's people, once again earn our sustenance and health by the "sweat of our brow." (Our work at The Land Institute in Salina, Kansas, is for the purpose of establishing a relationship with the landscape that is a bit closer to the relationship we had with ecosystems before we changed the face of the earth with extensive till agriculture. I do think we can make some inroads for coping with the Fall but mostly I want to deal with the second tier of problems.)

In considering the second tier of problems, let us accept till agriculture as a given. Even though it has been around only about five percent of our total evolutionary history, it has become so all pervasive that we now have no choice but to figure out how to manage it wisely. Let us look at the major problems in this middle or second tier.

1. Soil loss is greater now than it was fifty years ago, when President Franklin Roosevelt appointed Hugh Hammond Bennett to be

the founding chief of what was to become the Soil Conservation Service (SCS). It is not widely known that the loss of soil carbon is more serious than the loss of fossil fuel carbon through burning. The erosion loss of soil carbon and other soil nutrients to our offshore deltas and other places inaccessible to agriculture is more serious than the exhaustion of the metals and fossil minerals of our globe.

2. Soil loss lies at the core of the problem of agriculture. When the extractive economy of industry moved into the potentially renewable economy of agriculture—took it over, in fact—not only were the traditional problems of agriculture worsened, new problems were added. With the industrialization of agriculture the chemical industry made it possible to introduce chemicals into our fields with which our tissues had no evolutionary experience.

3. We now are almost totally reliant on finite fossil fuels for traction in our fields. It is in these second tier problems that we need to peel away the various masks, to assess the problems at a more fundamental level, and offer prescriptions for the culture at large. We will come back to this later.

The third tier of problems has to do with the work of numerous concerned people to make the best of a serious problem in the present. This involves the day-to-day struggle of helping farmers cope, helping them do the best they can. In a way, this work is like caring for a terminal cancer patient. We pour out our love and concern and help. We are with them through the period of chemotherapy or maybe surgery. We hope for immediate cures. We sit up at night. We suffer with them. We bury them. This is noble work, but it is more for the purpose of helping someone cope than it is to change the patient's fate. Sometimes to cope is to change, but we would be naive if we believed that we would cure the patient by easing the pain.

And so I return to the middle tier, with the hope that meaningful work in this tier can work to alleviate the first and third tier problems. If we are effective here, we might be able to soften the problem of the Fall. If we are effective here, many of the problems in the third tier may dry up entirely. Effective work in the middle tier, however, will involve a different way of thinking about our relationship to the earth. Some have called such a radical departure a paradigm shift. Whatever we call it, I think it is useful to understand, at least in a general sense, some of the history that has brought us to where we are.

I would like to begin with some Biblical history that pertains to the sin of idolatry—idol worship. As I understand that history, Hebrews were able to keep their monotheism alive because they were able to keep competing ideas of religious authority from corrupting their own loyalties. In the promised land of Canaan, the Hebrews confronted the baal worshipers, people who worshiped farm gods, gods responsible for every square foot of fertility. Every little village had its own baal. The old Hebrew god of the "mountain and the storm," the god that had brought them out of Egypt to Mount Sinai and finally into Canaan, had trouble competing with the baal gods of the Hebrews' Canaanite neighbors. The single, Hebrew tribe of goat and sheep herders in the rocky country to the south were better able to carry the notion of the god of the "mountain and storm" than those Hebrews who found themselves thrown into agriculture. These Hebrews' daily pattern was more like the pattern from their days of wandering in the desert. As farmers they had a terrible time resisting baal worship. Defeating baal worship meant defeating the idols that were made of the earth, in order to give emphasis to the more "correct" view of God, which is that He is a spirit. Clay and gold are materials; according to the Hebrew view, to pray to such "things of the earth" is to pray to the wrong stuff.

Such a worldview may have been essential for preserving the Hebrew people, whose lives were centered around the covenant with God, a covenant that was struck at Mt. Sinai where they "answered in one voice." But it created problems for European pantheists when the early Christians brought the legacy of their Hebrew tradition into Europe. Our pantheist forebears saw spirits in rocks, in waterfalls, in the deer of the forest, in the bear. By definition, Pan was everywhere. The early Christians who came into the wilds of Europe insisted that all of nature was "nothing but." To worship rocks and streams, bears and bees was to participate in the sin of idolatry. To lift our eyes up from the earth was culturally encouraged in another way, for even the most casual student of the stars could see there was order in heaven. On earth were uncertainties and constant problems with which we had to cope. The earth was an unlikely residence for God. Because the heavens were so orderly, any decent sort of god must be parked *there*. It was not that God couldn't and didn't roam around, but that heaven was, more or less, his permanent address.

For the Christian world, those who believed in a hereafter, the presence of God in heaven translated into an interpretation of heaven as their place of residence after death. These Christians viewed the earth as a launchpad, a place long on material and short on spirit. With the eventual extirpation of pantheism over vast stretches of the globe, the de-sacralization of nature was inevitable. The consequence is that science as we know it today was made possible. It is doubtful that the dissection of living animals and plants could be done by those who believe them to be holy. A pantheist would not view trees as so many board feet in the manner a Christian would. A pantheist would be less likely to measure the number of acre feet coming over a waterfall than his Christian descendant, centuries later, who had become a scientist. That which is sacred would be handled with a certain reverence.

I don't mean to pick on Christianity any more than any of the other major religions. Buddhists have often cut their sacred groves to build temples. Much of the Orient, where such religions flourish, has ruined environments. That is another subject, one that Yu-Fu Tuan has dealt with rather extensively. But our task is to understand our history a bit better.

Francis Bacon told us that knowledge is power, that the methodology of science would free us to organize the world sufficiently enough to give us a higher measure of comfort and security. More and more of us now know that comfort and security are not the solutions to the human condition, but few people knew it then. The experiment hadn't been run.

Long before the time of Bacon, people wanted power over nature. I don't doubt that there were some who believed that more power over nature would enable them to control their own lives. As

far back as the thirteenth century there were sporadic pockets of individuals who *were* breaking from the dominant circumstance in
which individuals' social positions determined their fates. Imagine,
a person's actions, indeed his very quality, were dictated mostly by
social position. This is hard for us to appreciate, since we own our
own labor power for sale in a competitive market. Nevertheless, we
are no more than sixteen generations away from a world that was
completely different. This change, which began in the thirteenth
century, did not culminate until the seventeenth and eighteenth centuries. The fossil fuel epoch and the opening of the New World co-
incided with the end of this era and the beginning of the Age of En-
lightenment and the scientific revolution. Power over nature, much
of it fossil fuel dependent, created lots of opportunity.

The question now becomes "So what?" The "sin of idolatry"
with its accompanying de-sacralization of nature paved the way for
the scientific revolution. After pantheism the world was more material than spiritual. Bacon was right. We see that knowledge is
power. We are in the fossil fuel epoch. In four or five hundred years,
we have inverted the structure of society from defining individuals
by social position to individuals determining their own social relationships. So we have had a bourgeois revolution. So what?

In their thoughtful book *The Dialectical Biologist* (Harvard University Press, 1985), Richard Levins and Richard Lewontin point out
that the social ideology of the bourgeois society, this recent invention, assumes that the individual is "ontologically prior to the social." By this they mean that individuals are free-moving social atoms with their own intrinsic properties. Society is a collection of
such individuals. In other words, society as a phenomenon consists
of the outcome of the individual activities of individual human
beings. This supports the view of Descartes, a view that became a
central notion of modern science. This view, this Cartesian view,
says that the part *has priority* over the whole. Cartesianism is not just
a tool or a method of investigation. It is a *commitment* to how things
really are. As Levins and Lewontin say, "The method is used because
it is regarded as isomorphic with the *actual structure* of causation. The

world is like the method." To say that knowledge is power, on the surface, may not sound all that bad. What was not perceived, I suspect, at the time of Bacon, is that the quantity of knowledge obtained by future scientific investigators would reward *them*, the investigators themselves, with power. "The success of the Cartesian method and the Cartesian view of Nature," Levins and Lewontin say, "is in part the result of a historical path of least resistance. Scientists work on the problems that yield to the attack." Investigators will not advance their careers, or should we say, they will not achieve power, by working on problems that they are unlikely to be able to solve. As Levins and Lewontin say, "brilliant careers are not built on persistent failure."

We can see readily how the path of least resistance *has* been employed in agricultural research. Practically no research has been devoted to the development of agricultural systems that will conserve soil, sponsor nitrogen fertility, manage water effectively, and control insects, pathogens, and weeds through biological, as opposed to industrial, means. Such research would require us to study whole systems and would violate the Cartesian view that places priority on parts over the whole.

So the question now becomes, "How do we break the stranglehold of Cartesianism?" Levins and Lewontin say that we should "look again at the concepts of part and whole." We used to justify holism or holistic thinking with the simple argument that the whole is greater than the sum of its parts. I know that I, at least, would nod knowingly and rest comfortably with such a simple justification. But Levins and Lewontin point out that "the parts acquire new properties . . . [and] as the parts acquire properties by being together, they impart to the whole new properties, which are reflected in changes in the parts, and so on. Parts and wholes evolve in consequence of their relationship, and the relationship itself evolves."

The purpose of the argument of Levins and Lewontin is to show that this relationship between parts and wholes, which is non-Cartesian, this relationship that has subject and object in constant interchange, this relationship of parts that can cause new properties to emerge in the parts themselves as the context changes, entails "properties of things that we call dialectical." That is to say, there is a thesis, an antithesis and a new synthesis or thesis. The Cartesian view believed that the world is like the method, that method was used because it is like the "actual structure of causation."

The authors point out how the Darwinian theory of evolution is a "quintessential product of the bourgeois intellectual revolution." First, it is a materialist theory in that it posits existing forces acting on real, existing objects, and so rejects the Platonic ideals. Second, evolution is a theory of change, as opposed to stasis. The nineteenth century was devoted to the idea of change, and biological evolution was simply a late example. Third, Darwin's idea of the adaptation of living things to the environment is, according to Levins and Lewontin, "pure Cartesian." The Darwinian assumption is that organisms change in response to an alien environment. The dialectical view accepts the first two premises of Darwin—the materialist theory and the theory of change—but rejects the third premise of Darwin— that organisms are *alienated* objects of external forces. The dialectical view holds that organism and environment interpenetrate so completely that both are at the same time subjects and objects of the historical process.

What is the utility of this history for those of us interested in achieving a proper relationship with the earth?

I have tried to show how the rejection of idol worship was rooted in the insistence that the material world is short on spirit, and that idols made from materials of this world, to the Hebrew and Christian mind, were not to be worshiped. This de-sacralization of nature helped set up the subject-object dualism. Darwin was a product of this culture, and though his theory of evolution involved real forces working on materials promoting change, he saw the environment, which is mostly physical, as consisting of objects that organisms had to adapt themselves to in order to live. Levins and Lewontin point out what numerous biologists and soil scientists have known for a long time: there is an interplay between organism and environment, and each is changed due to the presence of the other. Soil scientists are probably the most aware of this for they can readily see how the living world works to help form soil. Most biologists are less sophisticated.

Civilized people know that to objectify a person is dehumanizing, not only to the person but to the dehumanizer. Racism is a form of objectifying. Language that deals with the sexual parts of a person's body that does not carry a sense of reverence for the whole, we call obscene. Language that calls attention to skin color or ethnic background, elevating those factors above the whole person, we call racist.

We understand this very well when we talk about the human body but not when we think of nature. To talk about "the environment" as something apart from us is to separate us from the environment. We were, after all, made from the environment. We are maintained by it. The subject–object dualism has given us the notion that it is possible to isolate parts of the environment we don't like.

But there is more to consider before we turn this discussion back to agriculture and meeting the expectations of the land. Levins and Lewontin point out that many people will admit that social and economic factors strongly influence science. Newly graduated plant breeders with Ph.D.'s can command a starting salary one-third greater than newly graduated ecologists with Ph.D.'s. Why? Plant breeders can produce usable results faster than ecologists. Science is clearly influenced by the structure of social rewards and incentives. Look at the defense industry and its impact on science. But, as Levins and Lewontin point out, "nothing evokes as much hostility among intellectuals as the suggestion that *social forces influence or even dictate either the scientific method or the facts and theories of science.*" They believe that "science in all its senses, is a social process that both causes and is caused by social organization." Whether we like it or not, to be a scientist is "to be a social actor engaged in political activity." The speed of light may be the same under socialism or capitalism, but "is the cause of tuberculosis a bacillus or the capitalist exploitation of workers?" Would the death rate from cancer best be reduced "by studying oncogenes or by seizing control of the factories?" When

Monsanto produces seeds resistant to a Monsanto-marketed herbicide, "the environment" receives an increased herbicide load because the crop is "protected." Here is a clear example of placing priority on part over whole, and producing a social problem in the process. Denying the interpenetration of the scientific and the social is *itself* a political act. It allows scientists to hide behind scientific objectivity and, however unwittingly, to perpetuate elitism, dependency, and exploitation.

I think we are beginning to see some small measure of polarization in our universities on this subject. The professors and scientists who are most threatened by this little bit of consciousness-raising are the ones who, in the short run, can gain position and power in the universities by sweeping the problems resulting from Cartesian thinking under the rug. I see this as the source of the rise of that new caste of progressive fundamentalists, which includes the new geneticists. They may be highly trained but too many of them are poorly educated. The neural firings and intellectual pathways of the progressive fundamentalists are like those of religious fundamentalists. Most biotechnologists don't like the suggestion that molecular biology should be in a subordinate role; they are apparently uninterested in working as fellow scientists alongside ecologists. Fundamentalism is the product of a mind bent on power and uneasy with ambiguity. Fundamentalism begins where thought ends.

What does all of this mean for those of us who want to see the life sustaining resources available for the unborn, who have a sense of intergenerational justice, who have extended their love beyond the here and now? What does it mean for those of us who believe that farming is our most basic work? I think it means that we have to look at the interpenetration of part and whole and acknowledge that how we look at the world is how it becomes. I believe, for example, that there is a law of human ecology that, bluntly stated, is: "Values dictate genotype." I think we can safely say that our major crops, for example corn, soybeans, and wheat, have genes that we might call "Chicago Board of Trade genes." There are also wellhead genes and computer genes. In other words, there are ensembles of genes in our major crops that would not exist in their particular constellation were there not a Chicago Board of Trade (where a major share of the agricultural transactions occurs), or fossil fuel wellheads, or computers. Our values arrange even the molecules of heredity. That is interpenetration.

Gary Nabhan tells a story about a Native American woman in Mexico who had several ears of corn from her corn crop arranged before her as she shelled grain from each ear. There were ears that were tiny nubbins, and ears that were long; all had seeds of various colors. As she shelled grain from each ear to save for the next planting, Gary asked her why she saved seed for planting from the small ears. Her reply was that corn was a gift of the gods and to discriminate against the small in favor of the large would be to show a lack of appreciation for the gift. What she was doing, of course, was maintaining genetic diversity. Values dictate genotype. James B. Kendrick at the University of California at Berkeley says that if we had to rely on the genetic resources now available in the United States to minimize genetic vulnerability in the future, we would soon experience significant crop losses that would accelerate as time went by. Roughly one-third of our current crop comes from four inbred lines, which is roughly the same as the amount of variation that could be found in as few as two individuals.

I don't think that it is proper to say that the earth is an organism. An atom is an atom. A molecule is a molecule. A cell is a cell. A tissue is a tissue. An organ is an organ and an organism is an organism. Going up the hierarchy, we can say an ecosystem is an ecosystem and the earth is the earth. I believe that those who insist on calling the earth an organism are doing so because *they* happen to be organisms. We don't really know what the earth is, but we do know a little about it. We know that it is dynamic, that the inside is hot with heat left over from the earth's early days. And we have evidence that the hot core of our earth is responsible for life as we know it.

The old assumption was that the biota itself was self-renewing. Even in organic agriculture we assume that we can simply plant legumes, practice crop rotation, and thus renew a piece of land. This is true, but true in a sense that is more limited than we once believed. Within a very long time frame, a more accurate assumption is that the biota alone cannot rejuvenate an area; there must be some non-living capital (i.e., inorganic nutrients) that will accommodate life. In geological time, this capital is made available by large changes in the earth's surface, changes that are largely abiotic caused by glaciers, shifts of the tectonic plates, volcanoes. We know for example, that before the Andean uplift the Amazon flowed toward the west. Nutrients that were once headed one way are now headed another. In the pygmy forest of Mendocino County, California, there are terraces where each step represents about one hundred thousand years. We have evidence that the once verdant growth there has gone into decline, as nutrients have become unavailable over time. Land that once supported a lush redwood forest now supports a pygmy forest, vegetation that now appears to be greatly stressed. Yet life has been constant in this area. If life alone were enough, living forms—in this case, the trees of the pygmy forest—could bootstrap themselves to a level of greater diversity and larger biomass turnover, but apparently the necessary nutrients have leached from the soil and are no longer available for plant growth.

Life working alone on this earth is not enough. Reverence for life alone is incomplete. The pantheists were more right than they probably knew, for the very inner heat of our earth may be essential to make the geological moves necessary to sustain the biota as we know it. So are the gases, heated by the sun, that we call wind. Controlled by the moon, the tides provide a nutrient wash on our coasts to support an abundance of life. The interpenetration of moon and earth, of sun and earth, of soil and organism are all essential for our livelihood.

With the understanding that interpenetration is the right definition of our relationship to the earth, I come back to the crisis on the farm, which clearly is not an economic crisis. As I mentioned earlier, it is a crisis *reflected* by economic problems which derive from a larger, cultural crisis. It is not a crisis that can be cured by economics.

What is happening to the farmer and the farm is a faint foreshadowing of what is to come to the culture at large. The farmer is not an atomistic unit or satellite, sitting off to one side, needing repair. Neither is the farm. Agriculture in the largest sense cannot be repaired independently of culture and society. Vulnerability and helplessness begin with the fields, which are subject to erosion and pollution. Next most vulnerable and helpless are the people who work those fields. Next are the suppliers of inputs: the farm machinery companies, the companies that provide inputs, and the rural bankers. This is an inverted pyramid of vulnerability that begins with the farmer, and widens as we move upward to include the larger society.

The Cartesian worldview allows us to talk about trade-offs as though for each gain there must be only one loss. The ecological worldview, on the other hand, will tell us that one thing done wrong can create numerous problems throughout a system. Or stated positively, if something is done right, if something is done that fits, several problems are taken care of at once. The ecological worldview involves a profound awareness of the total interpenetration of parts.

Were we to act on the basis of ecological understanding, it would be possible to place the base of the pyramid firmly on the land. Society would then become a manifestation of what the land can support in a healthful and productive way. Only by meeting the land's expectations first can society be sustained. However, when we impose the industrial or extractive economy on the land, the base of the pyramid—representing society's wishes—is at the top. The point of the pyramid is stuck into the land like a hypodermic needle, injecting into the soil all the chemicals necessary to meet the demands of society.

The ecological pyramid illustrated in the basic ecology texts surely stands as a rough model for an alternative economic order. It has been billions of years in the making. In such an ecological economy the producers would be many and the mere consumers are few, exactly as Confucius prescribed for a healthy human society tens of centuries ago. Why have we inverted the pyramid? Cheap oil? Human nature? The oil, at any rate, is about gone; and never in the history of our country have we been more up against human nature than we are today. In 1776, this continent could absorb lots of bad human nature. The frontier was before us. Now, though our outward frontier has come to an end, instead of facing our problems squarely we keep looking to expand our frontiers inwardly, always for the purpose of exploitation. We have gone into the inner recesses of the atom and the nucleus of the cell. The exploitation of both atom and cell is not at all unlike ripping open the prairies, the very heart of our continent, or going into Third World countries like Brazil, where skilled welders are paid a dollar an hour to make farm machinery for America's fields. It is all of the same greed.

Frederick Jackson Turner developed the thesis that the American's self-definition is derived from the early frontier days, a time, we might say, of horizontal colonization. About the time we were fresh out of longitude and latitude we funded a space program and went for altitude. But colonization is not discovery. The quintessential aspect of colonization is exploitation and violence. Astronauts headed for orbit may be given more status than a farmer protecting a hillside from erosion, but a farmer who is successful in discovering ways to arrest nutrient loss on his sloping farm has made a more significant discovery than all the colonizers of space combined. So has the farmer who is gradually weaning himself from costly input farming, who is becoming less a consumer and more a producer.

Living Nets in
a New Prairie Sea

The Grass was the Country as the Water is the Sea.
WILLA CATHER

Author Joseph Kinsey Howard describes a spring day in 1883 in North Dakota when John Christiansen, a Scandinavian farmer, looked up while plowing a field to discover an old Sioux watching him. Silently the Sioux watched as the prairie grass was turned under. The farmer stopped the team, leaned against the plow handles, pushed his black Stetson back on his head, and rolled a cigarette. He watched amusedly as the Sioux knelt, thrust his fingers into the furrow, measured its depth, fingered the sod and the buried grass. Eventually the Sioux straightened up and looked at the immigrant. "Wrong side up," said the Sioux and went away.

Another writer in the mid-1930s described how his grandfather "broke prairie sod, driving five yoke of straining oxen, stopping every hour or so to hammer the iron ploughshare to a sharper edge. Some of the grass roots immemorial were as thick as his arm. 'It was like plowing through a heavy woven doormat,' grandfather said."

To many of us today it seems tragic that our ancestors should have so totally blasphemed the grasslands with their moldboards. But who among us, in their time, would have done otherwise? Nevertheless, it was one of the two or three worst atrocities committed by Americans that, for with the cutting of the roots—a sound that reminded one of a zipper being opened or closed—a new way of life opened, which simultaneously closed, probably forever, a long line of ecosystems stretching back thirty million years. Before the coming of the Europeans the prairie was a primitive wilderness, both beautiful and stern, a wilderness that had supported migrating water birds as well as bobolinks, prairie chickens, black-footed ferrets, and Native Americans. Never mind that the Europeans' crops would far outyield the old prairie for human purposes, at least in the short run. What is important is that the Sioux knew it was wrong, and that his words became regionally famous for the wrong reason. The story was often repeated precisely because farmer Christiansen, and the others who passed it on, thought it was amusing.

To their minds those words betrayed the ignorance of the poor Sioux. As far as the immigrant was concerned, "breaking the prairie" was his purpose in life.

Agriculture has changed the face of the land the world over. The old covering featured the top level of biological organization—the ecosystem. The new cover features the next level below, the population. For example, a piece of land that once featured a diverse ecosystem we call prairie is now covered with a single species population such as wheat, corn, or soybeans. A prairie is a polyculture. Our crops are usually grown in monocultures. The next most obvious fact is that the prairie features perennial plants while agriculture features annuals. For the prairie, at least, the key to this last condition resides in the roots. Though the aboveground parts of the prairie's perennials may die back each year, the roots are immortal. For whether those sun-cured leaves, passed over by the buffalo in the fall migration, go quickly in a lightning-started prairie fire or, as is more often the case, burn through the "slow, smokeless burning of decay," the roots hold fast what they have earned from rock and subsoil. Whichever way the top parts burn, the perennial roots will soon catch and save most of the briefly-free nutrients for a living future. And so an alliance of soil and perennial root, well-adapted to the task of blotting up a drenching rain, reincarnates last year's growth.

Soil still runs to the sea in nature's system, as in the beginning before land plants appeared, but gravity can't compete with the holding power of the living net and the nutrient recharge managed by nosing roots of dalea, pasqueflower, and bluestem. Banks will slip. Rivers continue to cut, as they did before agriculture, before humans. The Missouri was called "Big Muddy" before the prairies were plowed, a matter of possible confusion to those untutored by the river. But it is essential to realize that the sediment load before agriculture could not have exceeded the soil being created by the normal lowering of the riverbed, and what was carved from the interior highlands. It is even more important to appreciate that the amount of soil from the prairie that wound up in the river could not have exceeded what the prairie plant roots were extracting from parent rock or subsoil. Otherwise there would have been no soil over much of the watershed. What should concern us is the extra sediment load running in the river today—the fertility, the nutrients hard-earned by nature's myriad life-forms, which broke them free of their rocky prisons over the course of millennia, bathed them with chemicals, and made them fit for that freedom known only in the biota. The solar energy cost of mining these nutrients with root pumps is characterized by a slow payback period, an energy cost that only geologic time can justify.

Species diversity breeds dependable chemistry. This aboveground diversity has a multiple effect on the seldom-seen teeming diversity below. Bacteria, fungi, and invertebrates live out their lives reproducing by the power of sun-sponsored photons captured in the green molecular traps set above. If we could adjust our eyes to a power beyond that of the electron microscope, our minds would reel in a seemingly surrealistic universe of exchanging ions, where water molecules dominate and where colloidal clay plates are held in position by organic thread molecules important in a larger purpose, but regarded as just another meal by innumerable microscopic invertebrates. The action begins when roots decay and aboveground residues break down, and the released nutrients begin their downward tumble through soil catacombs to start all over again. And we who stand above in thoughtful examination, all the while smelling and rolling fresh dirt between our fingers and thumbs, distill these myriads of action into one concept—soil health or balance—and leave it at that.

Traditional agriculture coasted on the accumulated principal and interest hard-earned by nature's life-forms over those millions of years of adjustment to dryness, fire, and grinding ice. Modern agriculture coasts on the sunlight trapped by floras long extinct; we pump it, process it, transport it over the countryside as chemicals, and inject it into our wasting fields as chemotherapy. Then we watch the fields respond with an unsurpassed vigor, and we feel well informed on the subject of agronomics. That we can feed billions is less a sign of nature's renewable bounty and of our knowledge than a sign of her forgiveness and of our own discounting of the future. For how opposite could monoculture of annuals be from what nature prefers? Both the roots and the aboveground parts of annuals die every year; thus, throughout much of the calendar the mechanical grip on the soil must rely on death rather than life. Mechanical disturbance, powered by an ancient flora, imposed by a mined metal, may make weed control effective, but the farm far from weatherproof. In the course of it all, soil compacts, crumb structure declines, soil porosity decreases, and the wick effect for pulling moisture down diminishes.

Monoculture means a decline in the range of invertebrate and microbial forms. Microbial specialists with narrow enzyme systems make such specific demands that just any old crop won't do. We do manage some diversity through crop rotation, but from the point of view of various microbes, it is probably a poor substitute for the greater diversity that was always there on the prairie. Monoculture means that botanical and hence chemical diversity above ground is also absent. This invites epidemics of pathogens or epidemics of grazing by insect populations, which in monocultures spend most of their energy reproducing, eating, and growing. Insects are better controlled if they are forced to spend a good portion of their energy budget buzzing around hunting for the plants they evolved to eat among the many species in a polyculture.

Some of the activity of the virgin sod can be found in the human-managed fields, but plowing sharply reduced many of these soil qualities. Had too much been destroyed, of course, we would not have food today. But then who can say that our great-grandchildren will have it in 2080? It is hard to quantify exactly what happened when the heart of America was ripped open, but when the shear made its zipper-sound, the wisdom that the prairie had accumulated over millions of years was destroyed in favor of the simpler, human-directed system.

Where does all this leave us? Is there any possible return to a system that is at once self-renewing like the prairie or forest and yet capable of supporting the current and expanding human population? I think there is.

Much of our scientific knowledge and the narrow technical application of science has contributed to the modern agricultural problem. Nevertheless, because of advances in biology over the last half-century, I think we have the opportunity to develop a truly sustainable agriculture based on the polyculture of perennials. This would be an agriculture in which soil erosion is so small that it is detectable only by the most sophisticated equipment, an agriculture that is chemical-free or nearly so, and certainly an agriculture that is scarcely demanding of fossil fuel. We are fortunate in this country to have a large and sophisticated biological research establishment and the know-how to develop high-yielding, seed-producing polycultures out of some of our wild species.

At The Land Institute, we are working on the development of mixed perennial grain crops. We are interested in simulating the old prairie or in building domestic prairies for the future. Conventional agriculture, which features annuals in monoculture, is nearly opposite to the original prairie or forest, which features mixtures of perennials. If we could build domestic prairies we might be able one day to have high-yielding fields that are planted only once every twenty years or so. After the fields had been established, we would need only to harvest the crop, relying on species diversity to take care of insects, pathogens, and fertility.

This of course is not the entire answer to the total agricultural problem, much of which involves not only a different socioeconomic and political posture, but a religious dimension as well. But breeding new crops from native plants selected from nature's abundance and simulating the presettlement botanical complexity of a region should make it easier for us to solve many agricultural problems.

As civilizations have flourished, many upland landscapes that supported them have died, and desert and mudflat wastelands have developed. But civilizations have passed on accumulated knowledge, and we can say without exaggeration that these wastelands are the price paid for the accumulated knowledge. In our century this knowledge has restorative potential. The goal to develop a truly sustainable food supply could start a trend exactly opposite to that which we have followed on the globe since we stepped onto the agricultural treadmill some ten millennia ago.

Aldo Leopold lamented that "no living man will see the longgrass prairie, where a sea of prairie flowers lapped at the stirrups of the pioneer." Many share his lament, for what are left are prairie islands, far too small to be counted as a "sea." Essentially all this vast region, a million square miles, was turned under to make our Corn Belt and Breadbasket. But now the grandchildren of pioneers have the opportunity to establish a new sea of perennial prairie flowers, the product of accumulated scientific knowledge, their own cleverness, and the wisdom of the prairie.

Oracles, Prophets, and Modern Heroes

Writing in *Seven Tomorrows*, Paul Hawken, James Ogilvy, and Peter Schwartz underline our need to search for a middle road in our future. In their words, we need "a future that is neither so hopeful as to be unrealistic, nor so grim as to invite despair. Optimism and pessimism are not arguments. They are opposite forms of the same surrender to simplicity. Relieved of the burden of complex options with complicated consequences, both optimists and pessimists carry on without caring about the consequences of their actions. Convinced of a single course for the juggernaut of history, whether malignant or benign, both optimists and pessimists allow themselves irresponsible actions because they believe that individual actions have no significant consequences."

ORACLES AND TWO SPECIES OF PROPHETS

Before the era of linear programming models and the more complex systems approach such as the *Limits to Growth* study sponsored by the Club of Rome, we had other ways for dealing with the future. We listened to oracles and prophets.

What is essential about an oracular utterance is that it carries a truth that is not revealed as truth except in retrospect. When the event occurs, there always seems to be a twist that the prophet did not anticipate. Before Croesus went to make war against the Persians, he consulted the oracle at Delphi, who said that if he made war against the Persians, a great kingdom would fall. Croesus made war against the Persians and a great kingdom fell—his own. Those who believed in oracles lived in a world in which they saw human destiny as locked up in the stars, and existence as involving a certain inevitability beyond human control. They weren't hurt so much, therefore, by learning the truth until it was too late to do anything about it.

The ancient Hebrews had their "sayers," too—the prophets, and they were of two varieties. There were prophets like Jeremiah, who let the people know that catastrophe was coming, that nothing could be done about it, and so the people had better begin to figure out what they were going to do when things came apart. There was utility in such prophecy, for if one had to hit the road, as countless Hebrews did, one could at least decide which road to take, and how to pack. As Jeremiah predicted, Jerusalem fell, and the famous Diaspora of the Jews began.

The second type of prophet gave the people a chance to avoid certain problems if enough people changed and took specific actions. The Jonah story is perhaps the most famous illustration of this type of prophecy, probably because that big fish got into the act. Ninevah was a wicked city and Jonah's mission was to warn the people that they must straighten up or be destroyed. Jonah didn't believe they would straighten up and he set up a crude shelter on the edge of the city to watch them get whacked by the hand of God. But the people responded to his prophecy, and—much to Jonah's disappointment—the Lord spared the city. I suspect that Jonah was disappointed because he was wondering how credible he would be among the people of Ninevah. Skeptical minds would be at work, wondering if it was their changed behavior that caused the city to be spared or if it might have been spared anyway. Such a prophet is not a creature to envy. He is in much the same position as the scientist who must produce results.

I don't trust oracles and prophets absolutely, but then they are not supposed to be believed absolutely. For that matter, I am not convinced that today's mathematical prophets fare much better. They use matrices and show intersecting lines of such factors as population growth and resource depletion. With powerful computers they factor in several variables and plot the "limits to growth," as in the Club of Rome study already mentioned. The ancient oracles and prophets were often too general to be useful. The modern mathematical prophets are often too specific. Nevertheless, their predictions have great value, especially if they are mixed with the musings of those who insist that we need a more broadly defined bottom line in deciding what paths we should take to ensure a sustainable food supply for the future.

Our modern problem is that we are faced with what the authors of *Seven Tomorrows* called "complex options with complicated solutions." In an attempt to distill the complexity down to something manageable, I believe we can characterize two sets of minds at work. Imagine a pie with a small piece cut, but not removed from the plate. The small piece represents human cleverness, the large piece nature's wisdom. Imagine another pie cut like the first, but with the labels reversed. The first pie represents the agendas for the future of those who are devotees of the importance of human cleverness, with little regard for nature. The second pie represents the agendas of those who rely primarily on nature's wisdom.

The "human cleverness" folk are of a very different stripe from the "nature's wisdom" people. As I see it, the cultural battle to come has little to do with the traditional differences between Democrat and Republican, liberal and conservative. If we are lucky, it will be a conflict between the human cleverness folk and the nature's wisdom advocates. Of course, we have to exercise human cleverness *and* take advantage of nature's wisdom. But the problems will come as the culture works out the proper ratio between these two, as countless hopes, dreams, and *bona fide* needs both float against one another and bombard one another during the shakedown. Our cultural values will be paramount in determining the outcome of the conflict.

The future of agriculture is now being threatened by the human

cleverness camp. People in high places are having visions. Consider
the words of someone who must be a fundamentalist devotee of the
human cleverness school of thought. This is a scenario of an Illinois
farm of the future, offered in 1984 by Dr. John R. Campbell, Dean
of the College of Agriculture at the University of Illinois.

> *Scenario*: It is a clear June morning in Illinois. The computer awakens
> the farmer with music. . . . Information gathered and processed by
> computer during the night appears on the bedroom monitor. Sensors
> in nose rings and ear tags and implanted devices in farm animals have
> been scanned to ascertain their physiological state. Conditions are
> normal, except for a pregnant gilt whose breathing rate is increasing
> as she prepares to farrow her first 20-pig litter. Confined sows and
> cows coming into estrus have been identified automatically by sen-
> sors measuring electrical conductivity of vaginal secretions. The com-
> puter has already scheduled each for receipt of frozen embryos—30
> for sows, 4 for cattle; embryos containing high-growth, low-fat genes.
> A few will be miniature (low maintenance) fertile males programmed
> for precocious puberty. The automatic feed grinders and mixers have
> functioned satisfactorily during the night, all animals have been fed
> and watered, quantities of feed recorded as distributed, amounts con-
> sumed by each animal estimated and registered, remaining levels of
> grain, protein sources, and feed additives in bins and tanks measured,
> and replenished orders automatically placed with local suppliers via
> the computer network.
>
> After breakfast the farmer enters the field carrying a small elec-
> tronic device consisting of a 7 × 10 cm video screen, a small keyboard,
> and a notepad. Using the eraser end of his pencil, he types the word
> BUG. This command establishes communication with the farm com-
> puter, which accesses a program on a random access optical disk of
> the "juke-box"–like disk storage unit in the County Cooperative Ex-
> tension Office. The program leads the farmer through an identifica-
> tion and diagnostic decision tree using voice, text, and color photo-
> graphs. When the predicted degree of information has been verified
> by sampling, a least-cost, environmentally safe, biological control
> procedure is prescribed. The farmer types the word "order" and, be-
> fore leaving the field, hears the approaching helicopter of the local pest
> control specialist. . . .
>
> The children, Mary, 7, and John, 19, are in the home learning cen-
> ter. The computer is helping Mary increase her typing speed. Earlier,

by means of a simple, exciting game, it introduced her to the concepts of set theory. John completed many courses for the B.S. degree without leaving home. He obtained the courses, all on random access optical disk, through the computer network, and paid for them on monthly communication bills. They represent the most timely and best-taught courses from 13 universities on the network. . . .

Professor Campbell is not some isolated visionary. He represents a growing number of people who have an agenda for the future that is ever more demanding of human cleverness.

Writing in the *Los Angeles Times*, Larry Green is sufficiently euphoric to report that much of the research underway "may even lead to food production without either farmers or farms." He tells about "genetically engineered soil bacteria that produce an insecticide that kills soil insects. Plant seeds are coated with the Monsanto-developed bacteria that remain around the roots, protecting them from insects while the plant grows." He tells about the "hormones that increase milk production in dairy cows" and says that "at least one of these hormones could be commercially available in 1987, increasing the daily milk output of each treated cow by twelve to forty percent." No matter that overproduction is the main problem among dairymen now.

Green reports "efforts to breed super-sized farm animals by inserting growth genes from larger creatures into smaller ones." In the last few months, "scientists from the University of Pennsylvania, the University of Washington and the U.S. Department of Agriculture reported successfully injecting human growth hormones into fertilized eggs of rabbits, pigs and sheep. This experiment raises the potential of injecting elephant or sperm-whale growth hormone into embryo calves to eventually produce bovine giants."

The Office of Technology Assessment predicts we may have "10,000-pound cows . . . pigs twelve feet long and five feet high in the next decade or two."

Green tells of one research project "designed to produce food without plants, fields, tractors or farmers. It would create food in photosynthetic reactors, or giant leaf factories.

"The Rebeiz reactor would use bioengineered, photosynthetic membranes—or man-made leaves—to replace fields of plants. Instead of corn or cotton, the reactor would make glycerol." He quotes Rebeiz as saying, "With glycerol you can make anything. It's like having an oil well."

In early 1985, a scientist in the USDA, partially responsible for the new emphasis on bioengineering in agriculture, asked a rhetorical question of a small group of us assembled to worry about conservation in agriculture. He asked if we were interested in saving agriculture or in promoting a food-producing system. He made it clear that he was interested in promoting a food-producing *system*. Agriculture as the source of culture was not important to him. He was clearly in the camp of those who felt that we needed to continue, as one writer put it, to "hot-wire" the American landscape with commercial fertilizers, pesticides, and expensive industrial equipment, the kind that allows a quick linear passage over the agricultural landscape. There is a kind of heroism in all this. When we hear such talk it is often accompanied by a kind of righteous battle cry, such as "we must feed the world." Such heroic language comes from the industrial fundamentalists, those who assume that what is relevant for industry is relevant for agriculture. These researchers ignore basic ecology and pay scant attention to the fact that agriculture is absolutely wedded to the cycles of nature. They persist in this approach even though energy flows and everything that supports life operates in cycles—the circle is a good metaphor for biological nature. The industrial fundamentalist isn't satisfied just to square the circle, as agriculture has done for centuries, but sees agriculture more as a straight line of industrial mass production hitched to, and more or less parallel with, the extractive economy. The central dogma of the industrial fundamentalist is to *impose* the extractive industrial economy on the cyclic and renewing economy of nature.

I don't believe responsible behavior will result if we think of ourselves either as optimists or pessimists, for as the authors of *Seven Tomorrows* said, neither are arguments but rather opposite forms of the same surrender to simplicity. If we allow ourselves to be "convinced of a single course for the juggernaut of history, whether malignant or benign," we *will* allow ourselves irresponsible actions in the belief that "individual actions have no significant consequences."

What then is our starting point? We must remember that the primordial roots of science grew out of two opposite religious traditions: those who cast their lot with the astrologers, and those who sided with the magicians. Those who followed the astrologers assumed that our actions didn't matter, that our future and our fate were tied to the stars and the planets. The oracular tradition was part of this culture. The larger and more subtle truths of this tradition—that there are circumstances over which we have no control and that we should be humble before the Creator—were simplemindedly extended into realms of personal life where they had no business. Those who profited the most from this extension were fortune-tellers, or printers of cards and calendars with interpretations of the signs of the zodiac. Others abandoned their lives to the "juggernaut of history" and "surrendered to simplicity." At the point where the culture came to regard the oracle more as a source of cheap amazement than as a way of being made humble *before* the tremendous, the oracular utterance was more fit for a carnival than a temple.

The tradition that got the biggest boost in science, however, was that of the magician. The magician's idea was that the world can be manipulated, that creation can be rearranged, that our destiny, indeed our liberation, lies in the hands of humans. We can act. We can shape our destiny. The usefulness of this prophecy is that it provides us something to act on, either for the purpose of cutting our losses, as in the tradition of Jeremiah, or for correcting our course and preventing catastrophe, as in the tradition of Jonah.

I think these ancient traditions offer us a convenient taxonomy and that in working from them we can begin to sort our problems. There are some problems that are so immediate that we have no options that will allow us to divert the juggernaut. The best we can do is cope and hope. Prophecies in the tradition of Jeremiah are appropriate here. But we face other problems that we can solve. Here prophecies in the tradition of Jonah are appropriate, which if heeded could cause much of the countryside to be spared from impending disaster.

There are many modern prophets who may not even think of themselves as such. I can imagine a modern prophet saying, "If we employ bioengineering in order to unite cows and hogs in one organism, we will create monsters." I can imagine that scientists might carry out such bio-manipulation with impunity for decades, only to discover that they had created several monsters—the thousands of humans who saw nothing wrong with such wholesale manipulations of life. Such dehumanized monsters, devotees of human cleverness, have already been created by the biotechnology now available. These are people who see nothing wrong with breeding featherless chickens to cut butchering costs, or who have entertained the notion that the gene for eyelessness should be introduced into hogs in order to provide an economic gain for those raising hogs in close confinement. Blind hogs would be less nervous and would gain weight faster.

I can imagine a modern-day Jeremiah saying, "The techno-twits are invading agriculture. They are interested in a food-producing system that features electronic sensors in nose rings and ear tags and implanted devices in farm animals. They will bring you computers, bedroom monitors, frozen embryos, pituitary-accentuated beef animals, delignified, cellulose-digested wood pulp, and a computerized education for your children in a home learning center. Your children will learn set theory through an exciting electronic game and will earn their B.S. at home.

"Prepare your escape," the prophet will say. "You will watch the loss of your country at the hands of the exploiters as you watch millions of acres of American agricultural land wash and blow away, or be poisoned and mined for the extractive economy. Say no to this," the Jeremiahs will say, "study the Amish, find the other good examples. Find those who cast their lot more with nature's wisdom than with human cleverness. Think and work on the proper ratio in the pie. Form coalitions. Reward cooperation and discourage competition in your economics. But take heart: You are the proxies for the future. Avoid the center and the centralists of agriculture, those who love TV monitors more than Holsteins, computers more than fields. The peripheralists will prevail."

Prophets in the tradition of Jonah will be around too. "Work," they will say, "for legislation that supports rural community. Insist on legislation that supports the owner-operated landholding, where responsibility lies with the people who work the land, where the scale of the farm is related to what the operators can accomplish with their labor and intelligence. Take heart. Do not despair. Gird up your loins. Save the old culture. Save the country."

We do need a future we can affirm. We do need a future that is neither so hopeful as to be unrealistic, nor so grim as to invite despair. I don't believe there is a juggernaut of history, for there is plenty of evidence that individual and group actions have had significant consequences. We have it in our power to take the best of the old and promote an attractive and healthful rural landscape. We have in our scientific inventory, particularly in the areas of ecology, population biology, and evolutionary biology, an abundance of scientific knowledge derived from discoveries about ecosystems and how they work. That knowledge can be directed to great use in the development of a sustainable agriculture.

Farm Debt

The farmer and the farm, as a unit, stand between the voiceless environment and the vociferous public. Farmers are not exactly quiet, of course, but because they are so dispersed a minority, their raised voices, however loud, are ignored. If we were to look at the American farm and farm family the way we look through a prism that organizes light into bands, we would see most of the visible spectrum of all environmental problems. That few have made the connection between the farm and the environment as a whole is not surprising. To continue the analogy, the average student who looks at light through a prism has to be told that those bands of color are the glaring white light coming from the other side.

It doesn't matter that numerous farmers who have gone or will go bankrupt have had or still have millions of dollars worth of assets. If we were to do a proper accounting, nature has trillions upon trillions of dollars worth of assets and is in trouble. What if we had to pay the energy cost for the solar irrigation we call rain, for example? The potential for exploitation has always been the most lucrative where the assets are greatest. It doesn't take a computerized cost-benefit analysis to show that to roll a beggar is a waste of time.

Even as the exploitation of the environment and farmer is going full bore, we vote money to "preserve" the air and the water and we vote money to "preserve" the farmer and the farm. Our success at both is about what one would expect when we throw only money at a problem. The farmer and the farm *and* the environment are essential for our lives but taken for granted by the larger public. We love the farm and the environment, mostly, I suppose, because we are their children. We came out of both. Most of what we call the environment is what is left of the wilderness that we all came out of. Many of us are but a few generations off the farm. And so we love the environment and the farm and the farmer in the same way that we love the Native Americans. It is a form of condescension; a poorly masked way of despising our roots in the wilderness. We don't really want to live in wilderness except perhaps during vacation. But we do want clean air and water as pristine vestiges of wilderness. Most of us don't want to *live* on a farm either, though most of us would like to *reside* there. Most of us don't want to live with the Native Americans, or live the way they did prior to being on reservations.

Farm debt derives from society's attitude toward the farm and farmers. Because we take both for granted, farm debt is like environmental debt, except that with farm debt, when the farmer gets hurt directly, he can complain. But because farmers are so few and so dispersed, they are scarcely heard. The farmer and the farm, like "the environment," are looked upon, for example, as means to offset trade deficits. The farm is a place where we can externalize costs. The cost of pesticides to the farmer and the cost of pesticides to the soil and groundwater are regarded similarly by the public: "a serious problem that something ought to be done about." But the problem is more fundamental than this glib statement would indicate, for soil pollution is an *expense* of production. So are pesticides and nitrates in our farm wells. So is the loss of farmers from the land. Land prices, equipment prices, and fuel prices generate overdrafts when prices are low or yield is down. Talk within the smoked glass cubicles at the bank is serious then. Voices are low. (Now bankers are in trouble. A farm crisis develops when the banks are in trouble, not when the Russians back out of a grain deal.) An overdraft of groundwater deposited years ago brings less discussion in Washington than the national balance of payments deficit. The increased cost for deeper pumping will bring howls of protest, but the aquifer in decline can't protest. In the longer run, the overdrafts at the bank and the over-drafts of the aquifer are the same.

Nitrates in the water from the commercial feedlot and over-fertilized fields are harmful and even deadly to baby pigs and baby people. The well is tested. The well is shut down. But long before the farm couple is regarded as cranky or strident about their "bad luck," they give up the fight and simply scrape up enough money to buy into the rural water district subsidized by the Farmers Home Administration. The "new" water supply may already be showing signs of nitrate and pesticide pollution. Still nature is speechless.

The farm problem is not a financial crisis so much as a failure of culture. It will not be—cannot be—solved by a new farm program so long as the farm family is the primary locus for receiving money. The farm family cannot exist in any dignified way without farm community. Giving farmers federal subsidies is like giving Native Americans monthly government handouts as they muddle along on a reservation that is the epitome of a destroyed culture. The very ex-istence of such an abstraction as a reservation boundary has de-stroyed the chance for the return of native American culture. Today's reservations are as lethal to Native Americans as measles, smallpox epidemics, and cavalry charges were earlier. And so Native Ameri-cans live on subsidy, without dignity. The abstract wall creates de-pendency.

For today's farmers, the descendants of the white settlers who ruined Native American life, disaster takes the form of destruction of rural community by the industrial state. Of course, the farmer and the rural communities willingly bought into the industrial state, but in much the same way that Native Americans willingly traded for whiskey and smallpox-contaminated blankets. Temporary relief to farmers came in the form of legislatively altered depreciation sched-ules and tax breaks, so that this already overcapitalized segment of the society would continue to buy ever more equipment and other production inputs and keep equipment manufacturers solvent. For a long time, the farmer thought that this problem was his own fault, and to a large extent it was, and so he didn't complain even behind the smoked windows of the bank cubicle. Ironically, the hat he was wearing carried an agribusiness decal that advertised the fertilizer company or the seedhouse or the pesticide company or the farm machinery company partly responsible for putting him in debt. The decal is symbolically positioned to show who owns his frontal lobe. Whiskey and smallpox were faster.

Society is currently structured to accommodate the capitalist economy. This is why the Environmental Protection Agency cannot protect the environment, the Bureau of Indian Affairs cannot protect decultured Native Americans, and the USDA cannot protect farmers. I am not saying we should get rid of any of the bureaus. They may help the environment and Native Americans and farmers cope within the capitalist structure, but none will solve the problems they are charged with solving. If we were really serious about protecting the environment, the discharge pipes and stacks of industry would all plug directly into their intake side, and costs would not be externalized to a voiceless environment. If we were really serious about helping the farmer, we would treat agriculture as inherently biological and cultural, not industrial. We would see more crop rotation, more strip cropping; we would see animals on the farm rather than in large feedlots, their manure going onto the fields; and we would see more rural schools, rural churches, and rural baseball.

If the government is interested in continuing to subsidize agriculture, it should concentrate on supporting the farmer as part of a rural community, instead of passing money through the farmer to subsidize agricultural businesses. Without rural community, the money paid as a direct subsidy to the farmer quickly finds its way into the pockets of agribusiness. The government could pay the difference between the price of gas and groceries in the small communities and what farmers pay for both in the larger towns, thus keeping money circulating in the rural areas. But even that would be only a partial answer. Farm debt and ecological debt on the farm foreshadow what is to come for our entire culture and the environment as a whole, unless we change, and fast. For the farmer and the farm, problems are still multiplying, problems that had their genesis decades and even centuries ago. Most of the rest of the American culture, though, still living in the white light of affluence, is so dazzled by the brilliance that emanates from a high energy society that it is not yet able to see the full spectrum of environmental and economic problems. Until we begin to acknowledge that giving the green light to capitalism prevents us from really solving the problems, the environment will remain speechless, soil will erode, and farmers will remain broke, dispersed, and relatively quiet.

I've said bad things about capitalism, but I have just about as many bad things to say about many of the socialist arrangements in the world. The point is that we need a new economic order that respects biological and cultural diversity. Our current economic order is better designed to exploit bad situations than to alleviate them.

Agriculture is over-capitalized and farmers are in debt largely because the extractive or mining economy has moved to the fields. We need economic models which will account for the cycling of materials and handle the flow of energy (but not just any energy; I mean contemporary or solar energy) in a safe and orderly manner. This model can be found in nearly all natural ecosystems of the planet and is trustworthy because it was hard won in particular places over the globe during billions of years of evolution. Sometimes to cope is to change, but not often enough. We need to have in mind economic models of sustainability that are based in nature or in primitive cultures, so that proposals to help farmers cope with a bad situation can be evaluated against some standard of permanence.

Falsehoods of Farming

I have chosen this subject at this time of problems on the farm because if we are ever to find a solution to agricultural problems it will
come in bits and pieces from many quarters. I hope the title does not
mislead you into thinking that the problems on the farm are mostly
the farmers' fault. I don't believe that. Rather, I believe that the outside forces working against the farmer are far more responsible for
farm problems than what farmers are doing wrong. Farmers are
mostly powerless at the moment, but in the midst of our frustration,
we can do some things, however insignificant they may be at the out-
set. There is, after all, no reason that we should not begin to clean up
our part of the act. By working on smaller issues, we might even
improve the chance for the larger and much needed changes in the
structure of American agriculture generally.

And so I want to talk about some untruths that we have been telling ourselves for too long; untruths that keep us from seeing the
world more accurately and more as a whole than we are seeing it
now. What are these untruths and what can we do about them?

1. *Farmers will never get together to solve their problems because they are
too independent.* This is probably the worst falsehood told about us
and that we tell ourselves. I am not sure why we believe it and why
we tell it. Farmers are totally dependent on the oil companies, machinery companies, fertilizer and pesticide companies. Farmers are,
in one sense, helpless subjects of the corporate kingdoms of agri-
power. The lords of the manor in the feudal system of the Middle

Ages in Europe demanded no more of their subjects than modern suppliers of chemicals, machinery, and fuel demand of theirs. Agriculture has always been highly profitable, but not necessarily for those who work the land. The lord of the medieval manor was rich. The peasants who worked the land were not. The lords of corporate agribusiness who supply the inputs for industrial agriculture are rich. The farmers who work the land are not. The peasants identified themselves willingly with a particular feudal lord and declared their loyalty to that lord. Their modern counterparts who work the land wear hats advertising the corporate lords for whom they work.

Farmers *are* fairly closemouthed and many, I suppose, see this as an indication of their independence. But remember this: Subjects living under extreme oppression can also be closemouthed. This closemouthedness could just as well be regarded as the quiet of subjects, not of citizens. Subjects don't talk openly. Citizens do. Citizens discuss their problems more or less openly—not just their problems, but the problems of the times in general.

Now I want quickly to add that there may be good reasons why farmers are dependent and on the defensive. But to say that farmers are independent does not tell the true story and as long as we live under that illusion, we do ourselves a disservice. The feudal lords of corporate agribusiness would like to keep us believing that we are too independent to get together, for if they can keep us divided, they can keep us conquered.

2. *We have to get big if we are going to compete.* This is a particularly vicious untruth promoted in the last fifty years of farming. For decades it was the policy of the Farm Bureau and of government agencies. Some of its roots lie in the old Committee for Economic Development.

The last fifty years saw our country lose two-thirds of its farmers. In one year alone, thirty-seven thousand farmers went out of business. As of 1979 there were a scant two-and-one-third million farms in the United States; this group is increasingly powerless, politically. Many have blamed the USDA, agribusiness, the farm press and land grant colleges. In the minds of most, however, the economic problems that led to a loss of people from our countryside were the consequences of "natural" circumstances. Examples of "natural" problems include such well-rehearsed clichés as "farm units are too small," and "land is unproductive," and "some farmers are simply poor managers," and "too many well-intended farm programs don't work," and finally, "America's cheap food policy has been implemented at the expense of the food producers."

All of these standard reasons have *some* merit, and for most of my thinking years I accepted them as correct, if taken in the right mix. Since boyhood I have heard what I regarded as extremists at gatherings of farmers, whether at a sale barn or at a farmers' meeting, angrily blaming the demise of the family farm on some conspiracy. "Mature judgment" required that I dismiss most of this talk as a form of paranoia, lacking substance or fact. After all, no sufficiently large group of corporate managers could possibly get together to orchestrate such a cruel, un-American and altogether unhealthy social change, either for the evil purpose of lining the pockets of a managerial elite or—and many people probably think this—"for the good of the country."

A few organizations, however, working for a sustainable agriculture have begun to document this history, marshalling information that could well provide the basis for a strong grass roots movement for massive land reform legislation. Among the more notable examples is the powerful Committee for Economic Development (CED). Organized during World War II, the committee members were concerned with the mass unemployment they felt was sure to come after the war. This group became a strong advocate of removing people from farms. We must recall that the depression of the 1930s was still fresh in memory at the end of World War II. The economic problems of the thirties had fostered a great deal of social and political unrest; large numbers of people called our entire system of capitalism into question. Those who stood to lose the most, if that system should crumble, sought to avoid a repetition of such problems. It was in this context that the CED was formed. Part of the group consisted of corporate presidents known for their strong business sense, "experience in analyzing issues," and their usefulness in "developing recommendations to solve the economic problems that constantly arise in a dynamic and democratic society" ("An Alternative Program for Agriculture," CED committee statement, 1962).

Another group of CED participants were several university presidents. Early on, they explained that "through this business-academic partnership, CED endeavors to develop policy statements and other research products that commend themselves as guides to public and business policy: for use as texts in college economic and political science courses and in management training courses, for consideration and discussion by newspaper and magazine editors, columnists and commentators, and for distribution abroad to promote better understanding of the American economic system." They certainly had faith in the role of education, for they advocated an information "blitz" at several levels.

The CED suggested that the main problem was the persistent excess of resources in agriculture—particularly *labor*—relative to the new farm technologies. The men at CED had identified part of the problem as too many farm workers. They seemed puzzled that these people were reluctant to leave the farm. There must be something awfully attractive about the farm. However, these intellectuals and businessmen reckoned that it was the support of prices that had slowed the movement out of agriculture. We should not forget that these were men who were accustomed to making things happen. They wanted the exodus from the farm to be—in their words— "large scale, vigorous, and thorough-going." They proposed that the farm labor force, five years into the future, be no more than two-thirds as large as its then-current size of 5.5 million. "The program," they said, "would involve moving off the farm about two million of the present farm labor force, plus a number equal to the large part of the new entrants who would otherwise join the farm labor force in five years." In other words, get two million off the farm and keep farmers' sons from staying on the farm.

They had a plan for accomplishing this. Simply lower the guaranteed price supports on agricultural products, and economics would do the rest. The language of the committee is less blunt, which may indicate that even some of them were sickened by the project. But they had a mission, a job to do, and they explained that the basic adjustment required to solve the farm problem, the adjustment of resources (mostly people) used to produce farm goods, could not be expected to take place unless the price system were permitted to "signal to the farmers."

CED recommended that price supports for wheat, cotton, rice, food grains, and related crops be reduced immediately. They insisted that the importance of such price adjustments should not be under-estimated. Lower price levels would discourage further commitments of new productive resources to those crops unless they appeared profitable at lower prices. On this point the committee was most emphatic. They stressed that "for several reasons it is important that price supports be moved to levels that, if wrong, will be low rather than high . . . new resources (especially people) should be discouraged from entering agriculture, at least during the adjustment period, and the rate of entry in the longer run should not be excessive." These *leaders* apparently recognized that the sons of farmers would think that "if it is too tough for Dad to make a living here, why should I try?"

There were other effects that this blue ribbon group anticipated. Lower prices meant increased exports, and since many of them doubtlessly operated as middlemen in at least some of their enterprises, the price of the item was of less importance than the fact that they were there when the money changed hands.

Mark Ritchie has summarized how these powerful people must have viewed the primary benefits of their recommendations: (1) increased return on corporate investment in agriculture; (2) over two million farmers and families entering the urban labor pool, which would tend to depress wages; and (3) lower prices on agricultural products, which would increase foreign trade and provide cheaper raw materials for domestic food and fiber processors.

The collective opinion of those with direct money interests in the food system was that there were too many farmers. The communication throughout this establishment was complete, even though most members of the network may never have talked to one another. We have long known that the connections among the various strong corporate interests have included our top universities and Washington policymakers. Many of these people are part of a high level "good ol' boy" system loaded with "mature judgement." They share a *presumption* that they know what is good for the country—and, for that matter, the world—and from that assurance they wield their power to influence policy, usually subtly, sometimes not.

Of course many of them did care about poor people in foreign lands who needed American products. And when the government sought their wisdom they understandably advocated the kind of policy that was both familiar and had worked—for them. "If it is right for us," they seemed to say, "it must be right for the rest of the world."

Let's look at a concrete expression of this attitude. Remember that CED wanted their policy projected into the classrooms. The college text *Economic Development,* by Gerald Meier and Robert Baldwin (John Wiley and Sons, 1957), contains a description which reflects the "good ol' boy" presumptions. In a chapter entitled "General Requirements for Development," the authors describe how "economic criteria of investments may not be sufficient to bring about the necessary changes, and that some non-economical actions may be to invest in projects that break up village life by drawing people to centers of employment away from the village." The authors state that "new wants, new motivations, new ways of production, new institutions need to be created if national income is to rise more rapidly. Where there are religious obstacles to modern economic progress, the religion may have to be taken less seriously or its character altered."

Of course it is not true that bigness is better. How is it that among the Amish in northern Indiana, in northeastern Ohio and in Pennsylvania, the average farm is about eighty acres? As a group, these farmers are better off than the rest of us. It is not their size, but their diversity and low capital costs for equipment and their emphasis on community that make them prosperous. It isn't just their simpler life-style; they have *more* spendable income than their non-Amish neighbors. Their cost for inputs is less for they employ biology—human muscles, horse muscles, crop rotation—while we employ industry. In other words, the Amish are resilient because of a sufficiency of people, while the rest of us require good running equipment in order to be resilient. But this fast and efficient equipment is costly.

You have probably heard the story about the Amishman who had just purchased an eighty-acre farm. His neighbor asked him if he thought he could make it on these eighty acres. The new owner replied that he didn't know, but he knew he could make it on forty.

3. *One farmer feeds sixty-eight people plus you.* The number keeps changing upward. The statement is almost comical. One would think that anyone old enough to paint this sign and put it up on the interstate would know better. Perhaps it is tied to the falsehood of the independent farmer. It should say, of course, one farmer plus an awful lot of fossil fuel, plus John Deere and their counterparts, plus a lot of suppliers of all sorts of inputs and feeds. What doesn't get said is that as the number gets larger, more farmers have been put out of business. We should be outraged when that number gets larger.

4. *American agriculture is the most sophisticated agriculture ever to be invented.* I think I would put it the other way around—that it is probably the least sophisticated agriculture ever to be invented. This depends, of course, on what we are measuring. If one measures corn yields or wheat yields only, then there is no argument. If one measures the number of calories required to produce a calorie of corn, then the statement is untrue. If one measures the pounds of fossil water from the Ogallala required to produce one pound of feedlot beef then again the statement is untrue. If one counts the five bushels of soil lost to erosion to produce one bushel of corn in western Iowa, then the statement is false. I think there are more reasons to say that American agriculture is unsophisticated than to say that it is sophisticated, especially if we are interested in justice for future generations.

5. *We must feed the world.* First of all, when the cost of grain pro-

duction is high, the farmer is in no position to sell food cheaply or to give food away. Hungry people don't have money and neither do most of their governments. High capital investment for production keeps food out of the hungry mouths of the world. Though the average yields of the Amish are lower than their neighbors, they can afford to give more food away because of their lower production costs.

As I mentioned earlier, a more appropriate statement would be "The world must be fed." In many cases this means a more sensible agricultural structure in the needy countries. In eastern Africa, Malawi feeds its people well and exports lots of food. Their president has made it a policy that the wages of the people in the countryside must be equal to those in the city. So people stay on the land and have a healthy agriculture and keep agriculture healthy enough to produce food for export. Farther north in Africa, Ethiopia has a well-fed army, which it needs in order to prevent revolution. But people are leaving the countryside for the city. We all have seen the pictures of starvation in Ethiopia on television. Malawi's policy of equal income is probably not the whole story behind the differences between these countries, but it must be part of it. What if it were U.S. policy for rural wages to be as high as urban wages?

But this is a digression. The point is, we are not feeding the world, nor should we be. Over the long run, as much food as possible should be produced locally. Our food goes to those who can afford it. A good portion of our grain goes to Europeans, who feed it to livestock and thereby increase their meat consumption. We are not feeding the world; we can't afford to. Our costs of production are too high, so much so that our exports have begun to decline.

6. *We need to support the family farm.* Nearly all of the agricultural legislation written over the past few decades has purported to support the family farm. In spite of this stated intent, most of the legislation has the opposite effect. As a policy matter, it is, I believe, a bad idea to focus on the family farm. What we need to do is save and restore rural community so that the family farm is a product of rural community. As it now stands, the farmer launders government

money, called subsidies, passing it on to the suppliers of inputs and equipment. If we subsidize small communities so that consumer items cost no more there than in the discount stores or automobile dealerships of the larger towns, this should cause the money to roll over and over in the community before it finds its way to Kansas City and Chicago, Tokyo and Geneva.

As it stands now, it is not the farmer who is being subsidized so much as the lords of corporate agribusiness. The farmer just launders the money. Why shouldn't the profits be plowed back into the rural community instead of allowing the farmer and the farm, in the words of Maury Telleen, to become "a quarry to be mined"?

New Roots for
American Agriculture

Soil Conservation Service (SCS) officials recently set the annual average loss of soil on cropland nationwide at five tons per acre. A 1977 U.S. General Accounting Office report put the average annual loss at fifteen to sixteen tons per acre. An even earlier study, done at Iowa State University in 1972, estimated average annual soil loss to be about twelve tons per acre. Despite the disparity in these estimates, each makes one point clear: Soil loss continues at an unacceptable rate in nearly all of the areas tilled for agricultural purposes in the United States.

This failure on the part of conventional agriculture to develop a sustainable soil policy leads us to advocate development of a mixed-perennial, grain-producing agriculture on sloping soils. Such an agriculture would more closely reflect natural ecosystems, substituting for soil-wasting, petroleum-intensive annual monocultures. While these perennial mixtures would be derived from plants possessing little promise now for meeting human needs, there is every reason to think the scientific community has the know-how to develop a sustainable agriculture of this sort simply because of the advances in biology over the past half-century.

There are at least five important reasons why this nation needs to

pursue research into high-yielding mixtures of perennial grain crops
that eventually augment annual monocultures:

1. The array of soil conservation measures used across this land is
substantial, but the extra thought and effort necessary to make them
effective has not been and probably will not be very compelling.

2. The doctrine of enlightened self-interest assumes that the economic system is sensitive to long-term considerations (beyond one
person's lifetime), when in fact it is not.

3. Stewardship generally has failed.

4. Various forms of conservation tillage, particularly no-till, are
chemical-dependent.

5. The economic system is almost totally insensitive to ecological
necessity. Therefore, a reward-and-punishment system imposed on
farmers would have to be sufficiently powerful to accommodate
ecological necessity. This has not worked in the past and likely will
not work in the future.

TOWARD AN ECOLOGICAL AGRICULTURE

An ecological agriculture must consider human nature. Humans are
primarily seed eaters, particularly grass seed eaters. Even when eating meat, humans are but one step away from grass seeds, for in developed countries livestock rely heavily on grains. Any compelling
alternative to corn, wheat, and soybeans, therefore, cannot be a forage crop even if processed into pellets. If most of humanity is to stay
with its evolutionary heritage, we must have grains.

If the nation were to resort to high-yielding perennials to save its
soils, why not begin with the traditional annual crops and, through
wide crosses and subsequent breeding, convert them into perennials? Gene splicing between conventional and perennial crops might
also be investigated, but for now attention should probably focus on
wild perennials.

Traditional annual crops (corn, wheat, soybeans) have been genetically narrowed, and most depend heavily on fossil fuel for fertil-

ity and protection from competitors. Grown in monoculture, they lack anything close to the water management system of the prairie. This is important when one considers that some of the nitrogen used by the prairie is supplied by rain. Finally, farmers have selectively bred conventional crops over the millennia to respond well in monoculture. Less difficulty is likely in breeding perennial mixtures than annual monocultures, even if the parent stock consists of plants that historically have been regarded as useless or nearly so. Specialized germ plasm would be selected from the wild plants bred for high yield to mature in synchrony. There would be one or two harvests each year.

Three families of flowering plants would be involved in the perennial mixtures: grasses, legumes, and composites, particularly members of the sunflower tribe.

If one looks at the bottom line of yield alone, mixed perennial grain crops may be only marginally competitive, especially in the early stages of development. If, however, the nation were to allocate no more fossil fuel to conventional agriculture than to perennial polyculture, conventional agriculture probably could not compete.

Because the watchword is sustainability, mixed perennials will be less susceptible to environmental perturbations. The diversity of a polyculture, if it is like a natural ecosystem, will make the plants less vulnerable to insects and disease. The more efficient retention and use of soil moisture by perennial roots will make the ecosystem more resilient than that of a traditionally plowed field, with its high rate of soil runoff. This resiliency in turn contributes to yield because reduced soil loss means retained fertility even after severe rainstorms.

Most perennial polycultures presumably would be grown first on marginal land. The availability of such land is enormous, more than 100 million acres. These grain mixtures could be safely planted on most of this acreage. Though yields would be lower on this poorer quality land, the penalty to the farmer would be much less in the initial stages than if these mixtures were planted on the currently cultivated landscape.

The economics of incorporating this biological approach to agriculture into what is fundamentally an industrial-style agriculture should not be prohibitive. The cost for seedbed preparation and planting should then be the same as pasture establishment now. If mixed grains grown on marginal land become grain sources for livestock, this amounts to an economic plus for farmers accustomed to feeding traditional grains.

All inputs would be greatly reduced in perennial mixtures, including fertilizers, pesticides, labor, and energy. These reductions, in turn, would reduce the farmer's capital investment and dependency on credit.

During the course of polyculture research and development, plant breeders at various institutions, particularly the land grant universities, would be assisted by their colleagues in agricultural engineering in adapting to conventional combines to harvest the experimental plots. In short, no fundamental technological innovation requiring huge capital investment seems necessary. Technological changes will likely be modest.

Annual polyculture was once common in underdeveloped countries. The classic corn-bean mixes are probably the best known. Perennial polyculture in which fruit seeds are harvested has been limited to woody vegetation (e.g., fruit and nut trees). Herbaceous perennial seed-producing mixtures in which seeds have matured at the same time would be harvested in a substantially different way than the polyculture of annuals or trees and shrubs. The latter do not lend themselves to machine harvest, but the herbaceous perennial polyculture does. Furthermore, some, if not most, of these perennial mixes would not have to be grown in rows.

In recent years there has been a steady increase in reduced tillage or no-till farming. A prime reason has been to reduce soil loss. Perennial polycultures would have the same result. But from here on, the differences between the two methods nearly overshadow the similarities. No-till is pesticide-dependent. Perennial polyculture would be essentially independent of pesticides. No-till features annuals in monoculture. Perennial polycultures imitate natural ecosystems.

PROSPECTS OF POLYCULTURE

In the long run, breeding high-yielding perennials may be the least difficult problem for developing a sustainable agriculture. There is the problem of achieving the right species mix for a particular region. Different farmers have different preferences and needs. One piece of ground is not like another. Even with all these vagaries, polyculture should make as much sense as monoculture. In a literature review on annuals in polyculture, D. C. L. Kass concluded that polyculture is beneficial, especially if there is a good choice of crops in conjunction with other environmental variables. Getting the right mix appears critical. More specifically, Kass concluded that the benefits of polyculture are clear in terms of nutrient withdrawal from the soil and economic return. When one of the crops is a legume, there is an overall improvement in the nitrogen status of the soil-plant system. All this leads, Kass contends, to greater stability in yields over time. This should be true of perennial polycultures as well.

Monitoring and husbanding a biological mixture may at first seem unacceptable to farmers because such an agriculture appears complex, so complex that experts are necessary. No farmer wants or needs the constant presence of experts to ensure a profit. But the problem of management would be less than one might expect. Range management on a per acre basis requires nothing close to the thought and effort that goes into a corn or wheat field.

Harlan commented that "the general principles that apply to the dynamics of natural grasslands apply just as well to the dynamic balance of species in tame pastures." Likewise, management of perennial polycultures of grain crops would apply many of these same principles.

YIELDS OF SELECT PERENNIALS

The incentive to develop high seed yields in perennials has been small compared with the effort that has gone into development of high-yield annuals. Native seed merchants have probably had more reason than most to increase yield, but even for them a modest increase in germination has been more important than a yield increase. The SCS regional plant materials centers have likewise had reason to increase the yield of various perennials, and they have done so in several species. But they have little time to devote to such work. They are woefully understaffed, considering their mission of supplying seed for the nation's three thousand conservation districts. Range agronomists have understandably paid close attention to forage yield increases, devoting only minor effort to seed-yield improvement. To date, there has been nothing close to what one might consider an aggressive national program for increasing herbaceous perennial seed yields.

Are herbaceous perennialism and high seed yield mutually exclusive? Woody perennials probably can produce high seed yields because of their greater height and exposure to light. But what about herbaceous species? Part of the answer depends upon what is meant by "high yield."

Winter wheat is regarded as a high-yielding crop. As an arbitrary standard, assume then a yield of 1,800 pounds of winter wheat per acre (thirty bushels at sixty pounds/bushel). Subsequent adjustments can be made for protein, carbohydrates, and oil yields on a per acre basis.

A survey of the literature on herbaceous perennial yields revealed the following: Buffalograss (*Buchloe dactyloides*), which had been fertilized and irrigated, yielded 1,727 pounds per acre. Although this yield included the burs, of which seeds are a small part, the yield represents fruit/seed material. Fertilized and irrigated Alta fescue (*Festuca arundinacea*) averaged 1,460 pounds per acre. A native stand of sand dropseed (*Sporobolus cryptandrus*), under dryland conditions at Hays, Kansas, yielded 900 pounds per acre.

Irrigated legume yields were also encouraging. A five-year average for Illinois bundle flower (*Desmanthus illionensis*) amounted to 1,189 pounds per acre. Fertilized cicer milkvetch (*Astragulus cicer*) yielded 1,000 pounds per acre, as did sanfoin (*Onobrychis viciaefolia*), which received seventy pounds of nitrogen.

Two species of sunflower produced exceptional yields when irrigated. The gray-headed coneflower (*Ratibida pinnata*) yielded 1,600 pounds per acre. Maximillian sunflower (*Helianthus maximillianus*) yielded 1,300 pounds per acre.

While these high yields from relatively unselected plants are encouraging, there are cases in which yield may be somewhat less important than the quantity of protein per acre. Based on the per acre yield mentioned for Illinois bundle flower, cicer milkvetch, and sanfoin and their respective protein content, all have a protein yield of about 400 pounds per acre. This exceeds the protein from an acre of wheat yielding 31.6 bushels in which the protein, by weight, amounts to 228 pounds. It even compares favorably with 100-bushel-per-acre corn, which yields about 500 pounds of protein per acre.

As more data on perennial seed yields is collected, numerous perennials with high-yield potential may be discovered. For example, almost all grass seed yields are from grasses ordinarily used for forage and hay. Such grasses have been selected to send much of their energy to the leaves rather than to the seed. Perhaps there are perennial grasses that are poor for forage but good for seed yield that have not been studied, such as the dropseed genus *Sporobolus*.

The direction of future perennial grain crop research may depend upon the results of research yet to be done, which could determine which perennials should be grown in rows and which in solid stands. Because prevention of soil erosion is the prime motivation for doing and promoting perennial grain crop research, erosion studies are needed to determine the relative ability of perennials in rows or solid stands to control erosion. Rows of perennials may not reduce soil erosion to a sustainable level. In that case, seed yields of perennials in solid stands will need to be recorded in order to find candidates for further yield improvement.

Although the selection program to improve seed yields of herbaceous perennials is practically nil compared with the extensive breeding done on annual crops, improvement in perennial yields will almost surely be more dramatic than from a comparable effort with annuals. So little work has been done with perennials that there is enormous room for development. These gains might be expected to come from the same sources breeders historically have called upon to increase yield, by reallocating the resources within the perennial plant and by increasing plant performance overall.

In deciding whether a particular species is a good candidate to be a good seed producer or not, a species should demonstrate high variation so the breeder has a basis for selection; show promise as a potentially high seed-yielder unless it serves another role, such as being a strong nitrogen-fixer or natural repellent of insects and pathogens for associated species; suggest some potential for determinate seed set, shatter resistance, and a favorable reproduction-to-vegetation-shoot ratio; promise a relatively high yield for a minimum of three years in order to accommodate itself in a three-year or more replant cycle; and exhibit meiotical stability, that is, the ability to display normal chromosome behavior during sex cell formation. For example, if the species is of polyploid or hybrid origin, the probability of regular meiosis should be very high or show some promise of high regularity. With perennials, meiosis need not be as regular as with annuals.

BENEFITS OF PERENNIAL POLYCULTURES

An agriculture based on a perennial polyculture would produce a number of dramatic benefits for society and the nation's natural resource base. The resource-oriented benefits include the following:

1. Soil loss would be cut to zero, then soil would begin to accumulate. The fossil energy savings for fertilizer to replace this lost soil would be significant.

2. The indirect consumption of fossil energy in agriculture would decline substantially. For example, the application of commercial fertilizer would drop greatly because, like a prairie, a perennial polyculture would use nitrogen more efficiently.

3. The efficiency of water use and water conservation by the perennial ecosystem would be near maximum, and springs, long since dry or short-lived during the year, would return. Irrigation would decline, reducing fossil energy use.

4. Commercial pesticides, especially those with no close chemical

relatives in nature, would be abandoned or nearly so. In addition to the reduction in chemical contamination, a minor fossil-energy savings would result.

5. The direct consumption of fossil energy at the field level would be greatly reduced.

Among the social benefits of perennial polyculture would be the following:

1. Long before such an agriculture would be completely in place, its efficient use of water would render more manageable major policy questions surrounding irrigation projects that involve aquifer mining or water diversions that lead to soil salting and silting problems.

2. Policy considerations that include equitable land ownership rather than major corporate control of land would become more manageable due to the reduced need for high capitalization in machinery, energy, farm chemicals, irrigation, and seed.

3. More than one hundred million acres, currently considered marginal land largely because of the serious erosion potential under current cropping systems, could be brought into production, reducing the pressure for ever-increasing land prices. Stable land prices would increase the opportunity for people to have a farm as a place to promote and experience right livelihood rather than as a food factory.

SOIL SAVINGS

The primary purpose for developing such a radically new agriculture is to save soil from being eroded or chemically contaminated. What would be the effect on soil conservation if herbaceous perennials occupied the 316 million acres in the United States currently planted with the top ten crops? Results vary, of course, depending upon the estimate of soil loss chosen and the number of years between replantings. Let us assume that all 316 million acres, including the cotton and hay acreage, were planted to mixed perennials. By replanting fields every five years, 253 million acres of the 316 million acres in any given year would remain unplowed. On the basis of a soil loss of five tons per acre, that would save 1.27 billion tons of soil. Given a twelve-ton-per-acre soil loss, the savings would exceed three billion tons a year, an amount equal to the annual loss when SCS was established.

These estimates are conservative. They are not adjusted to account for the soil buildup during the years of no plowing, nor are they adjusted to reflect that plowed perennial fields, if they behave like plowed perennial pastures, would lose less soil the first year than a field that has experienced continuous annual plowing.

It is also possible to calculate the energy required to replace the nitrogen, phosphorus, and potassium in the eroded soil on these 316 million acres. If that loss is five tons per acre, the energy value is about 96.4 million wellhead barrels; at twelve tons, the energy value is 231 million barrels; and at 15 tons, 289 million barrels.

The energy value of the nitrogen, phosphorus, and potassium lost through the harvest of 100 bushels of grain from an acre roughly equals the energy lost in nine tons of eroded soil on an acre—more than half a barrel of petroleum at the wellhead in each case. Assuming a five-year replant cycle, the savings could be equivalent to 80 million wellhead barrels if the loss is five tons per acre or 239 million barrels if the loss is fifteen tons per acre.

Substantial energy savings are also possible as a result of reduced seedbed preparation and field work, pesticide and fertilizer use, and irrigation.

PROSPECTS FOR RESEARCH

Two constraints impinge on the prospect for development of perennial polycultures. The first constraint is that promoters of conventional agriculture believe the traditional methods of soil conservation—terracing, grass waterways, check dams, contour farming, stubble mulching, conservation tillage, and crop rotation—are adequate for the task. Many apologists for conventional agriculture believe that where these methods are not practiced it is either the result of a farmer's ignorance or bad character. At the deepest level of contemplation, soil loss is regarded as a lack of exercise of stewardship. However, any exercise of stewardship that employs these traditional measures may be beyond the ethical stretch for *most* farmers. By this I mean that they may be unable to practice what they know to be correct behavior. Good practices require attention, care, small scale, intelligence, and economic solvency.

The second constraint is the consequence of the somewhat uncritical acceptance of conservation tillage, which is expensive and chemical-dependent, for growing traditional row crops, particularly corn. This acceptance indicates that most agriculturists believe a technical fix is already at hand to reduce soil loss significantly.

What can be expected of agribusiness and government in the way of research on perennial polycultures? Little help will likely come from private seed companies. Who can blame them for not producing a perennial or groups of perennials that could put them out of business? To leave them out of the action, however, may be unfortunate. These companies have important resources, particularly human power for extensive analysis, an absolute necessity if new crops are to be developed. But if these companies are to be involved, they will likely want in return patents on their products.

The agribusiness firms that sell fertilizers and pesticides would likewise appear to be poor candidates for promoting cropping systems that reduce the sales of their products. Involving them requires a stretch of the imagination. Perhaps instead of paying them for applying "medicine," some of which has harmful side effects, they might be paid for contributing to the maintenance of soil health and balance. Nutrients will have to be added, but why should the nutrient and pesticide bill be based on quantity alone? It is in the spirit of the recent emphasis on integrated pest management to recognize that a species mix is automatically a chemical mix, and if that chemical mix discourages insect and pathogen spread, the organizations and individuals responsible could be rewarded. The entire idea is somewhat like paying a doctor a fee while the individual is healthy, with payments ceasing or slowing when an individual is sick. In our increasingly service-oriented economy, "package plans" are sold more and more.

This is not a promotion of agribusiness involvement. The burden of defining their role in the development of a sustainable agriculture should be on their shoulders, not on the shoulders of the rest of the people in a free market economy. But neither should private seed companies or other agribusiness firms be written off before they are given a chance to show how they can participate in developing a sustainable agriculture.

The USDA has the greatest potential of any institution for defining the mission of a sustainable agriculture. If it decided internally that such an agriculture were appropriate, the machinery exists within the department to give sustainable agriculture the necessary emphasis and financial support.

Private research agencies perform tasks very well when they are clearly defined. They could perform many of the biochemical studies and much of the cytogenetic screening.

Probably the most competent of all would be the national laboratories, though the nature of their work does not lend itself to the type of research considered here. The most ideal situation would be the establishment for sustainable agriculture of a mission-oriented agency comparable to the National Aeronautics and Space Administration.

A TIME TO ACT

Government traditionally has responded to the direct interests and needs of farmers so long as there is some indication of ever-increasing yields. Research has been oriented to high yield even when that has meant the demise of small farmers and the growth of corporate farms and agribusiness. When one considers that food exports are a major way of softening our mounting balance of payments deficit, it seems unlikely that the emphasis will change. In 1981, the United States sold about $45 billion worth of agricultural products, which pales in comparison to the $100 billion-plus bill for foreign oil. This history, in which the emphasis has been on the bottom line, is partly the consequence of the fact that the time frame of too many politicians has been "until the next election." One might hope, however, that leaders would see the long-term economic sense of a truly sustainable agriculture and allocate substantial funding to the development of perennial polycultures.

Future policy considerations will be critical to the adoption of this new agriculture. Agribusiness could be threatened. These economic interests will have good reason to push for policies that give traditional crops an edge over the perennials. Inequities could come in the guise of concern for fair treatment. For example, government subsidy of energy across the board for all farmers, for whatever reasons (assured food supply, sympathy for their economic straits) could slow the adoption of perennial polyculture. The cornfield, in such a system of subsidy, does not have its higher energy cost subtracted and thereby gains unfair advantage as a competitor against the lower-yielding, but less energy-expensive perennials.

With no subsidy, however, any particular field could compete in dollar return after adjustments are made for lower energy input; lower fertilizer application; reduced pesticide use; lower equipment cost, due to reduced machinery use in seedbed preparation, planting, and cultivation; reduced soil loss; and reduced labor.

This article, coauthored by Wes Jackson and Marty Bender, originally appeared in the *Journal of Soil and Water Conservation.*

A Search for the Unifying Concept for Sustainable Agriculture

The use of "*the*"—instead of "*a*"—"unifying concept" in the title of this chapter was not accidental. I chose it because I believe that a truly sustainable agriculture will be directly keyed to nature, which already has a well-understood unifying concept of its own. I am speaking of the unifying concept of biology as discovered by Darwin, which explained how new life-forms came into being through natural selection. This was later coupled with the discovery that the DNA-RNA hereditary code is universal. Essentially all the natural ecosystems of the earth are many times more complex than our most elaborate agricultural ecosystems, and if there can be a unifying principle for the diverse natural biota, why shouldn't there be one for agriculture, especially if nature is our standard?

Some could argue, of course, that agriculture is so much a product of human manipulation that a unifying concept will have to be *invented*. I don't think so. I think it is only to be *discovered*. There are enough examples of good farming that we recognize as being in harmony with nature that we can look to them. By good farming, I mean practices that don't consume ecological capital or otherwise degrade the landscape. That such examples are few in number doesn't matter. Other possibilities for agriculture that are theoretically possible, although yet to be developed, take nature's designs even more into account than the best agricultural examples now available.

This search did not begin with this chapter. It is at least as old as the Hebrew scriptures and has been getting more complicated ever since. Before the industrial age, soil loss alone was at the core of the "problem of agriculture." It is currently the oldest of our ecological problems, for it probably has its origin with the first tilled crop laid out on a sloping landscape. The industrial revolution added other problems to the core—fossil-fuel dependency for traction and fertility and chemical dependency for pest control. The problem becomes even more complicated once we add the structural changes that have especially influenced agriculture in the developed world as the result of new technologies interacting with politics and economics.

THE HIERARCHY OF STRUCTURE

As problems of production agriculture become more complicated, it was probably natural that we would begin to look for a way of thinking about the definition and possibility of sustainable agriculture, a conceptual tool that would provide a perspective on all crops and all management practices. First of all, we need a way of thinking about any agricultural unit: the family garden, the truck farm, the small family farm, the large corporate farms, and the individual field. That is one thing a unifying concept would do. It would be fluid and become the basis for a taxonomy even though taxonomic schemes are not fluid. Concepts, like movies, deal with process. Biological classification, at least, is expressed as single frames, stopped in time. A taxonomic scheme can be totally arbitrary. One could classify agricultural efforts, for example, based on acreage or principal crops or livestock grown, but the utility of such a scheme would be limited unless certain laws about "the nature of things" stood behind the classification system.

Professional ecologists do research and talk about the possibilities and limits of natural diversity. Why should we not be able to deal with farm ecosystems in a similar manner? What if population biologists considered how populations of organisms interact with other populations on the farm as well as with the physical and chemical world? If the taxonomy took into account these considerations and the laws that govern the successful establishment of any new wild species, the classification would surely be regarded as more natural *and* more complete. The concept would provide the basis for a taxonomy by making us more conscious of the variables that must be adjusted if agriculture is to be sustainable in any given place. The best taxonomic scheme for our purposes, then, is more than an arbitrary system of classification, for it would be embedded in a theory of structure.

I don't mean to imply that we will ever have a perfect taxonomy that shows precisely where and how everything fits. That would be asking too much. Our current classification of the biota is only an approximation of the phylogenetic relationships in the tree of life, and a once-and-for-all classification of nature's life-forms based on phylogeny still eludes us. But just as Darwin's unifying concept for biology—evolution through natural selection—made a natural classification of our earth's biota theoretically possible, so might a unifying concept for sustainable agriculture make a classification of sustainable agriculture methods *theoretically* possible even though it may be practically very difficult.

In his paper on the ecosystem as a conceptual tool in managing natural resources, Arnold Schultz explains why ecosystems should be studied as a separate science. He thinks the study of ecosystems should be a separate field because it is not like studying botany, zoology, not even ecology. Schultz argues that such a science would provide a framework for analyzing any organization integrated immediately above the level of the individual organism.[1]

The ecologist A. G. Tansley, who coined the term *ecosystem* in 1935, described it as "the whole system, including not only the organism complex, but also the whole complex of physical factors forming what we call the environment."² It would be more accurate to say, "forming what we normally have been calling the environment of the organisms." Defining *ecosystem* more succinctly than Tansley, R. L. Lindeman described it as "a system composed of physical-chemical-biological processes active within a space-time unit of any magnitude."³

Good descriptions are difficult. Both of these could, I suppose, include the earth, sun, and moon, and so they are limited. Rather than worry for now how large an ecosystem can be before it is no longer an ecosystem, let us consider where it stands in relation to everything else downward in the universe. J. S. Rowe says the ecosystem is part of an integrated hierarchy of things from atoms to molecules to cells to tissues to organs to organisms.⁴ The ecosystem stands immediately above the organism. That the ecosystem should stand at this particular place is not immediately obvious because beyond organisms we are used to thinking about plant communities, biotic communities, vegetation, populations, species, and world fauna and flora. Schultz asked this question and finally agreed with Rowe that what an ecosystem has that these categories do not have is "thinghood." Rowe's argument is that nature has "chunks of space-time or 'events' which have both qualitative and quantitative properties. Events which endure are known as objects." For something to have "thinghood"—that is, to be an object—it must have volume because volume is the basic component of perception. These other categories, species, populations, and so on, lack volume because they do not include their surroundings.

But volume alone won't do because, as Schultz says (leaning on Rowe), "for objects to have a high degree of 'thinghood' they must exist in both space and time. Form and function must be constant or have rhythmic stability. Organized entities have strongly marked structure and function. When organized entities have strongly marked structural and functional characteristics, they are *perceived as autonomous* and stand out as natural objects of study" (italics mine). The integrative level of atom, molecule, cellular organelles, cell, tissue, organ, organ systems, and organisms are all natural objects of study. They are "slabs of space-time," as Rowe calls them. The volumetric criterion holds for them all. Before ecosystem was added to the hierarchy, J. K. Feibleman wrote twelve laws of integrative levels.[5] With this volumetric consideration, Rowe rewrote one of Feibleman's laws so that volume and space relationships are included. This law holds that *the object of study, at any level whatever, must contain, in the volume sense, the objects of the lower level, and must itself be a volumetric part of the levels above.* Each object on any given level constitutes the immediate environment (in the sense of impinging surroundings) of objects on the level below. Each object is a specific structural and functional part of the object at the level above. The ecosystem passes this test as an object because it consists of individual organisms plus the nonliving world that connects them. (What the next natural "chunk" or "slab" beyond ecosystem in nature is, I can't say. Perhaps it is bioregion or continent or ecosphere.) What is important for agriculturists is that Rowe seems to have made the case that the next integrative level above organisms is the ecosystem as the space-time unit. To repeat, species, population, vegetation, or community are not "environment" for individual plants, and they are not any specific volumetric functional part of the ecosystem.

The ecosystem does differ from the other categories in the hierarchy because the human defines the boundary. The boundary of an organism is natural and well understood. The same is true of an organ or a cell. There are certain natural ecosystems, such as bogs, in which the boundary is clear. However, for most natural ecosystems, it is difficult to know, with much precision, where the boundary lines are. Particularly problematic is knowing where tall-grass prairie ends and mid-grass prairie begins or where mid-grass prairie ends and short-grass prairie begins. Yet few prairie ecologists would deny the existence of the three prairie types.

With agricultural ecosystems, where we place the boundary is more suggested by a human-imposed pattern on the landscape than by nature. A deed can define the boundary of a farm. A fence line may define a pasture or a field. We could think about a farm community as an ecosystem or a farm or an alfalfa field or even a cubic meter of soil. We can place mental cubes around anything we want because our purpose is to be better accountants of what goes in and out through the boundary and, at the same time, appreciate the dynamics of the ecosystem as a structure that obeys certain laws common to the other levels in the hierarchy.

LAWS OF THE INTEGRATIVE LEVELS

Before we dwell on the ecosystem level as an integrative level, we need to consider the other laws of integrative levels distilled by Schultz to apply to the ecosystem. Schultz has also derived these other laws and their corollaries from Feibleman's twelve laws. Next, we can't ignore species and populations if we are to talk about agriculture because we can't ignore populations of cows or wheat. Even though species and populations did not attain the status of being part of the hierarchy of structure that Feibleman, Rowe, and Schultz have discovered, they were regarded seriously as candidates. They just failed the test of "thinghood." They are, nevertheless, part of another kind of hierarchy with its own laws. Species and populations exhibit some dynamic properties fundamentally different from what individual organisms do. They generate species diversity, and experience adaptation. Species adaption is of a different order than individual adaptation. Perhaps knowledge of these dynamic properties and the ways in which they impinge on the ecosystem will be useful in the effort to discover the unifying concept for sustainable agriculture. But first let us look at the other laws of integrative levels that Arnold Schultz, writing with the ecosystem in mind, developed from J. K. Feibleman's scheme.

Schultz combined some of the laws and corollaries of Feibleman into seven categories that he felt were the most relevant to the development of the ecosystem concept. I have already mentioned the seventh law, which deals with the criteria for "thinghood" and makes the ecosystem an integrative level. The first six laws follow, almost as Schultz has stated them, but not in the same order. I am listing the self-evident laws first. They are not necessarily trivial, but an elaboration on their relevance to agricultural ecosystems may not be necessary at this time.

1. *In any organization, the higher level depends on the lower.* Just as the organism depends on organs, organs on tissues, tissues on cells, and so on, so the ecosystem depends on organisms, soil, and water. What is implied here is that the lower the level is, the more enduring the level is. Atoms are more lasting than molecules. The physical and biological components of an ecosystem are more lasting than the ecosystem.

2. *The higher the level, the smaller the population of instances.* There are fewer molecules than atoms and fewer ecosystems than organisms. The levels form a population pyramid. This law still holds for agricultural ecosystems. The variety of farms in Kansas may be greater than the number of natural ecosystems, but the number of total ecosystems still does not exceed the number of organisms associated with the farms.

3. *Complexity of the levels increases upward.* Complexity is partly the result of accumulating structure, but most of the complexity stems from emergent qualities that pile up. The emergent qualities are the interrelationships that increase exponentially while the number of components increase linearly.

4. *Each level organizes the next level below and adds emergent qualities* (see number 3). If one knows only the properties of the lower level, the emergent qualities are unpredictable. Schultz points out that we couldn't have known from the gases hydrogen and oxygen that they could produce water. In this example more than one water molecule had to exist before the liquid property could emerge.

There are examples at other levels of organization where "critical mass" is necessary before an emergent quality can arise. For example, one cell does not a tissue make. This "critical mass" idea may have great practical importance for sustainable agriculture. There is probably no such thing as a completely sustainable farm anywhere in the United States. Some farms are more resilient or can stand being weaned away from the fossil-fuel economy better than others. The Amish farms are probably the most notable, as a group, in this respect. But it is rare that an Amishman or Amish family will venture forth, alone, into a locality not previously settled by Amish people. When colonies are founded, seven families usually go together. The Amish know they can't make it in isolation. Although there are probably a lot of isolated good farmers whose farms experience no net soil loss, their ancestors were probably heavily dependent on community during the farms' establishment. A lot of back-to-the-landers learned that most individuals or even individual families who have made attempts at sustainable agriculture have fared poorly or failed outright. Perhaps a "critical mass" is necessary before the emergent qualities necessary for a sustainable agriculture begin to appear. For our purposes, the implication is that sustainable agriculture will need rural communities if it is to survive and flourish. In retrospect, it is clear that, once the systematic destruction or dismantling of rural communities was underway in the United States, the weakening of the family farm was inevitable.

5. *For an organization at any given level, its mechanism lies at the level below and its purpose at the level above.* This law may need to be restated. As it now stands, it presents a particularly sticky problem when thinking about nature. Ironically, it serves us well in thinking about conventional agricultural ecosystems even though it does not apply to natural ecosystems. We may know the purpose of a cornfield, but what is the purpose of tall-grass prairie? Our problem becomes especially difficult if we don't keep the origin of a structure separate from what we finally see as a finished product. For example, if we drop down to the organ level and take the kidney, we might say that the purpose of the kidney is to remove nitrogenous wastes from the blood of an organism—to cleanse it. Here the purpose lies above the organ (at the organism level) and the mechanism lies below the organ (in the tissue). But few biologists would be satisfied with such an explanation, for in the evolution of the vertebrate kidney, few of them would believe that purpose was "pulling" on one end, forcing the development of a mechanism to accommodate a higher purpose. Biologists, instead, speak of adaptation and say that those early creatures that had tissues fortuitously tilted in the direction of removing

nitrogenous wastes, however slight, were positively selected, in a Darwinian sense. Creatures that carried improvements on each former adaptation were further selected, and so on.

Only from our vantage point in history do we assign the words *purpose* and *mechanism* for what nature has produced. Even if we had been observers in the early stages of the evolution of the kidney, we probably would have considered the primitive or elementary kidney a finished product.

Perhaps part of the human condition is the result of the distance from nature in which we place ourselves by dealing with the world in a language that emphasizes purpose and mechanism. An example from agriculture is a typical Kansas wheat field. If it is truly representative of a field on a typical Kansas farm, its purpose is to provide cash outright for the farmer. The bottom line, in other words, is production. The farmer needs a high yield. Since his emphasis is on production, he naturally employs mechanisms of "mass production," products of industry. Mass production features a huge capitalization of equipment and inputs and seeks to minimize labor costs. The logical outcome is larger and more expensive machinery and fewer people. But this creates vulnerability. If the crop is to get planted, tended, and harvested, breakdowns of equipment become less tolerable. For many farmers, this often means new equipment every two or three years and a need for higher yields to pay for it. The *purpose* of the usual American farmer, to produce cash, almost to the exclusion of everything else, *dictates the mechanism* for growing wheat.

On 160 acres recently purchased by The Land Institute, there is a clear example of the impact on the land from the purposes of production agriculture. On our quarter are two small streams that run during part of the year but are usually dry by the middle or end of summer. This quarter section had been held in a trust by a local bank for a young man for twenty-five years and had been rented out to one farmer for the past seventeen years. As a farmer who derives his total income from farming, he is typical for our area. He tills around twelve hundred acres of land and has a cattle operation, too. Because these two streams and their woody growth proved to be such a nuisance for his large equipment during the dry period when it is time to prepare the ground to plant wheat, he and a trust officer at the bank, probably with government funds, had the riparian community bulldozed out so that there was one field that he could "farm right through." He could lower his production costs if he wasn't slowed down by small fields. Tree roots would rob some nutrients from the wheat along the edge. Most years it it still too wet in June to cut the wheat that grows in this wet spot, but it does not slow him down during seedbed preparation and planting in late summer and early fall. Ironically, when a large limb of a big hackberry growing on the property line fell directly into the field a few years ago, he farmed around the limb. Because he did not have to reduce the speed of his tractor or make a wide turn to farm around the limb, it was, because of his purpose, not cost effective to remove it. Neither was it cost effective to till the ground on the contour or build terraces. It was cost effective to pull into the field and start going around the perimeter of the field and move toward the middle, letting the perimeter, not the streams, dictate the pattern. The living riparian community was bulldozed out, the dead limb was allowed to encroach on the field, and the natural topography was ignored in tilling and planting, all for the same purpose.

I recently visited several Amish farms in Ohio where the topography is much more rugged than in central Kansas. Streams divide their fields, too, but their farms have a very different appearance. What is underway on an Amish farm does not involve single purpose. The farms are not regarded as economic units, although the Amish make sound economic decisions. What we observe on the Amish farms is similar to what we observe on a natural ecosystem— homeostasis. Purpose and mechanism are transcended.

One Amishman told me that he liked the long-stemmed wheat because the straw was about as important to him as the grain. He used the straw for bedding by which he salvaged and improved his manure, and he was surprised to hear that we Kansans would either plow the straw under or burn it. The Amish feed some of the grain to livestock, some they sell. Although they take the Biblical injunction to "dress and keep the earth" seriously, one does not sense that they go about their daily work with that phrase rolling around in their heads. They are interested in profit and high yield, but neither concern drives them as a singular purpose. Had The Land Institute's newly acquired 160 acres been an Amish farm, it would have been highly diversified, and provided there was a surrounding community of Amish, that 160 acres would easily have supported one Amish family. The living riparian community on each side of the two streams would have been a habitat for an abundance of wild species including quail, pheasant, and deer.[6] It would have been a source of fuel, a boundary dividing the farm into smaller fields. It would break the hot, dry wind of summer, interrupt insect migration, and host some predatory birds and insects. The smaller fields would have suited a horse- or mule-powered agriculture. The large cottonwoods would have provided shade for grazing animals or for a resting team and driver. The fallen hackberry limb would have been converted into firewood. The straw that we plow under or burn would have become bedding for livestock and thus become a way of holding urine and manure, and all three would have been returned to the fields from which they came. Some of the grain would be fed on the farm, some would be sold, depending on need.

Because the emphasis for the Amish is not exclusively on production, mass production of food on the farm is incompatible with their sense of how to live in the world. To "dress and keep the earth" is biology. Mass production invites mechanism. Mechanism and machines go together. That is why conventional farmers are obligated to assume the high capital costs of expensive equipment. To ensure that the crops are planted, protected, and harvested, the Amish depend on a sufficiency of people who live on the land and do the work. For the Amish, *resilience* lies with that sufficiency of people. It isn't likely that during harvest all the workers will catch the summer flu at the same time. Furthermore, the Amish, while always busy, are not pressed for production; if a draft animal does become sick, harvest isn't shut down. Wheat is cut by a binder and shocked in the field to dry; eventually it is loaded onto wagons drawn by horses or mules, carried to the barn, and threshed in the barn by belt power provided by a stationary tractor. The straw is blown to the place where it will be stored until it is forked into a stall for bedding; the grain is blown into a bin. In the Amish community, on each farm and in each field, purpose and mechanism are subordinate to the larger goal of homeostasis.

6. *It is impossible to reduce the higher level to the lower.* Since each level has its own characteristic structure and emergent qualities, reduction is impossible without losing those qualities. There is no greater reality in the parts than in the whole; they are equally real. This concept is widely overlooked or ignored by most scientists today. In 1972, P. W. Anderson wrote a paper entitled, "More Is Different."[7] He emphasized that each level of organization is more than layers of atoms, that each level has its own laws every bit as fundamental as the fundamentals of physics. Nevertheless, we have been taught that many of the commonsense observations all around us are an illusion and that the component parts are the reality. It is like saying that sympathy (an emergent quality) is an illusion, since it cannot be located by the dissection knife.

When we look at the major problems of modern agriculture— soil loss, chemical dependence, increasing dependence on fossil fuel, loss of the family farm, more corporate farms, an expansion of agribusiness—we see that most of them are the consequence of too much reductionism. It is understandable. We live in a society dominated by scientific reductionism. The problem is deep and probably resides in the history of science and technology. Both have been deeply influenced by physics, which we placed foremost in the hierarchy because we believed that once we understood the building blocks of nature, all else would be chemistry. Consequently, the most acceptable explanations in science have been in physical rather than biological terms. One can see where this leaves the biologists. They were simply discouraged from postulating scientific laws for the various biological levels because scientists in general regard biological phenomena as too indeterminate for safe prediction.

This justification for and emphasis on reductionism in science and technology in general has been carried over from physics to biology to agricultural science. At the extreme, it appears that a license has been issued to an agricultural establishment of scientists and technologists to function as salesmen of industrial farm inputs. These salesmen of everything from fertilizer to computers push their products with an abundance of quantitative documentation of their performance. They define problems for farmers—problems farmers scarcely knew they had—and then sell them the remedies. On the slickest of paper, these salesmen may display impressive tables and graphs, products of highly mathematical econometric models developed and expanded by econometricians in the Department of Agricultural Economics back at the land grant university. In some respects, these academics are worse for the farmer than the salesmen, for they function as the agricultural priests who pass on the fundamentalist faith in reductionistic thinking.

Meanwhile, many farmers, along with the rest of society, have come to distrust their commonsense observations. They have been made to feel that they are an illusion. Fewer and fewer farmers think about their farms as an ecosystem. A good farmer will continue to look at a particular hillside and see what possibilities it offers in the total scheme of things, which includes his farm as a whole, its history, his family, and the aptitude of everyone in the family. An agricultural economist usually does not consider any of this because he, along with most of society, still distrusts many commonsense observations.

We have also been warned away from analysis at higher levels of organization because, as we move up the hierarchy of the sciences from physics to biology, it appears that various attributes become so complex and variable that, at best, experts in systems analysis are necessary to keep track of them all. But as Eugene Odum has said, "It is an often overlooked fact that other attributes become less complex and less variable as we go from the small to the large unit." Odum reminds us that there are "homeostatic mechanisms, that is, checks and balances, forces and counterforces, [which] operate all along the line, and a certain amount of integration occurs as smaller units function within larger units."⁸ The rate of photosynthesis of a forest community, in an example Odum provided, is less variable than that of individual leaves or trees within the community because when one part slows down, another may speed up to compensate. Odum went on to say that "when we consider the unique characteristics which develop at each level, there is no reason to suppose that any level is more difficult or any easier to study quantitatively."

As we move up the hierarchy, "more is different," partly because emergent qualities develop at each level. Although the findings in the study of one level may be useful in understanding the next level above, they never completely explain the next level. To refer to an earlier example, in water all the attributes of the higher level (the liquid quality of water) were not predictable by knowing only the properties of the lower level (the gases of hydrogen and oxygen).

The ecosystem, as a volumetric "thing," can be a useful conceptual tool when thinking about sustainable agriculture. We can put mental boundaries around whatever part of the landscape we want to examine. A good farmer is constantly making mental cubes or spheres that include the vertical as well as the horizontal dimensions of a field or farm or even a farm community. To call them cubes or spheres oversimplifies what a good farmer does. He is at once an accountant or acknowledger of what passes through the various boundaries. If he simplifies his farm by selling off the livestock, for example, he has simplified more than his work. Mostly because of the nature of the economic system, he has simplified his thought, a kind of tragedy, for what farms need now, desperately so, is more thought by their owners. But just as important, perhaps, is being able to acknowledge that the ecosystem obeys the same laws as the other integrative levels.

THE OTHER HIERARCHY

As useful as the ecosystem concept may be for those of us thinking about sustainable agriculture, it is not enough. As suggested earlier, we cannot think about agriculture without thinking about *popula-*

tions or *species*—cows, hogs, sheep, chickens, wheat, corn, rice, and so on. Just because these domestic species and their thousands of populations failed the test of "thinghood," their importance is not diminished. So we need to talk about species and populations that fall under another hierarchy, one that may have its own rules or laws. Although they are yet to be discovered, one would expect them to be independent of the laws for the hierarchy of structure.

It is useful and accurate to think of each of the numerous species operating in a prairie—be they song birds or grasses or bacteria—as *boundaries of biological information.* Prairie chickens don't mate with upland plovers or scissor-tailed flycatchers, let alone with big bluestem. The gene pools of the different species are mostly bounded. Plants representing 237 species have been counted on a square mile of native prairie in Nebraska.[9] When we add to those the species of vertebrates, algae, fungi, bacteria, actinomycetes, annelids, mollusks, centipedes, millipedes, insects, and arachnids, the quantity of biological information is awesome. As part of this hierarchy, every species fits somewhere in a phylogenetic tree, as part of a phylogenetic hierarchy. Birds are more related to one another than they are to reptiles, and they are "higher" on the evolutionary scale because they came from reptiles, not the other way around. Anyone who has thought about it is struck by the range in size and shapes of all life-forms from the viruses to the whales and redwoods.

There is another kind of diversity, the diversity within a species. Although our human viewpoint may be such that we may scarcely notice the variations within a population of meadowlarks on the prairie, meadowlarks do vary. So does a population of compass plants and so do populations of all the other species. Some are more noticeable and more variable than others. The variation in dogs, for example, is dramatic.

This range of variation within a species has something to do with the remarkable extremes we find in species dispersion. Humans, of course, are the most cosmopolitan of all and are highly variable. On the other hand, fellow creatures on this planet are noteworthy because they are restricted in their range; there is probably very little variation within their populations. For example, a fungus species called *Labaulbenia* grows exclusively on the back part of the elytra of a beetle. The beetle is endemic to limestone caves in southern France. There are fly larvae that develop exclusively in seepages of crude oil in California. Larvae of a certain fruit fly species develop only in the nephric grooves beneath the flaps of the third maxilliped of a land crab endemic to certain islands in the Caribbean.

It is not uncommon to discover that genetic variation is on the increase within a species whether or not that species is expanding its range—that is, the biological information within the boundary of that species is increasing. Two biologists, D. R. Brooks and E. O. Wiley, have postulated that at a certain point when biological information is accumulating within a species, when new genes are being added, the information of that species is becoming more chaotic, more entropic.[10] It seems to follow that the flow of energy through such a species also becomes more chaotic or disordered. Brooks and Wiley believe that, under certain circumstances, this disorder, which they call high information entropy, actually causes populations to divide or speciate. First of all, whether Brooks and Wiley are right or not as to cause, it seems appropriate to assume that each of the two new populations has less information to contend with than does the parent population from which they were derived. (Even if we divide a herd of Holstein cows in half, there is less variation in each of the two resultant populations than in the original population.)

With speciation, the ecosystem is usually a beneficiary in that its energy flow is more ordered, less chaotic; new niches are formed, new relationships are established. The ecosystem may become more stable. But even species diversity within an ecosystem has its limits. There can be too much. For years, natural preservationists and ecologists alike believed that more diversity within an ecosystem meant more stability. This is not necessarily true. R. M. May has written a book that summarizes the numerous examples showing that diversity and stability do not necessarily go together.[11] Many of his examples are of simple communities that were more stable than similar but more complex communities. In other words, information entropy can exist within the bounds of an ecosystem just as information entropy can exist within the bounds of a species. If Brooks and Wiley are right in their contention that information entropy within species is responsible for the evolution of new species, the possibility follows that information entropy at the ecosystem level is responsible for the evolution of new kinds of ecosystems.

It requires but little imagination to realize that a particular farm can have too many species. If there are too many kinds of animals, there will be too many chores to do. There can be too many crops to tend. But in our time, too many species on a farm is seldom the problem; the usual problem is too few.

To think only about a high information agriculture, meaning lots of species diversity on the farm, is not enough. We have to think also about taking advantage of the natural integrities that exist in all the diversity. In a polyculture of plants, one species that fixes nitrogen complements others that are unable to fix nitrogen. Another species may do a better job of pulling up trace elements necessary for the nitrogen fixer and other species as well. This is diversity with complementarity; much, but not all, of it is fortuitous. On a diverse farm, we see complementarity. Some of the corn fed to cows passes through with the manure and is eaten by hogs and chickens. Both the cow and hog manure contain meals for chickens. With so many vitamins and minerals, bright orange yolks loaded with barnyard energy and nutrients fall into a skillet and contribute to the health and quality of life for the human. Here diversity of species slows entropy but complementarity is as important as diversity. It is important to remember, however, that without diversity we lack the basis of complementarity. Without manure from the cattle to enrich the land that supplies feed for the chickens, the human farmers could not have complementarity. Natural integrities are dependent upon diversity.

Up to now, we have considered two hierarchies in our attempt to discover the unifying concept for sustainable agriculture. The first is the hierarchy of objects that includes life-forms. We have seen that there are certain common laws for the various integrative levels in this hierarchy, from the atom to the ecosystem. They are very general laws but useful, nevertheless. The second is the hierarchy of ascent from the first cells to modern mammals, flowering plants, and so on. This hierarchy is the consequence of speciation, and with speciation, information entropy is reduced. Whether information entropy is responsible for speciation is still an open question among biologists. For our purposes, it is important to consider only that division of an information package reduces information entropy and probably energy entropy as well. In other words, for biological information to be used in an optimum manner, new boundaries have to be established because, at some point, a restriction of information is more necessary for stability than a free flow of information.

Agriculture comes out of nature, out of the two hierarchies described here: hierarchies that intertwine but obey certain laws independently. To think about agriculture is to think about life, and to think about life is to think about the way nature has bounded its information to live on the earth and why it is so packaged. What are the rules? Are any of them universal up and down the hierarchy? Are they as universal as the laws of integrative levels of structure? If we don't know the laws or rules, is it possible that some of our problems with agriculture are the direct result of this ignorance?

THREE FACTORS COMMON TO BOTH HIERARCHIES: SCALE, INFORMATION, ENERGY

One factor that appears to be of importance in a consideration of both hierarchies is *scale*. In the hierarchy of structure, scale increases upward from atom to organism. Each level has natural boundaries, so each level is unambiguous. At the ecosystem level this changes because humans draw the boundary lines. As scale increases both upward through the hierarchy or at only one level, whether it is a cell, an organism, a field, farm, or farm community, *a linear increase in size is seldom if ever attended by a linear increase in relationships*. An ostrich egg as a large cell necessarily has different properties from a hummingbird egg. For purposes of illustration, a better example is found among potatoes grown by the Peruvian Indians high in the Andes.[12]

These potatoes are generally smaller than "improved" or commercial varieties, and estimates of the number of natural varieties range from "well over four hundred" to more than two thousand. In spite of what experts may think, to increase the size of the native potatoes could be a serious mistake. Larger potatoes would be harder to cook at the high elevations where the people farm and live and they would be more watery; additionally, since the skin contains so many nutrients, an increase in size would greatly change the surface to volume ratio, so that the potatoes would yield proportionally more starch than nutrients.

The size of the individual potatoes involves considerations formally understood by the laws or rules associated with the hierarchy of structure. The number of genetic varieties numbering from four hundred to two thousand involves considerations formally understood in our thinking about the hierarchy of descent. Scale is an important factor in both hierarchies.

Farmers assume the responsibility to manage an ecosystem, an enormous conceptual task partly because the boundary of the ecosystem, unlike the other levels in the hierarchy up through organisms, changes with scale. It is somewhat arbitrary where we draw the lines for a human-managed ecosystem, for it depends on what we wish to measure or to observe going in and out through the boundaries, whether it is a garden, field, pasture, farm, or farm community.

When speaking of ecosystems, we are talking about a community of organisms and the physical world that connects them. We are also talking about all the spaces between as components, not as void, because the actions between boundaries occur in those spaces. The ability of space to expand is another reason why scale considerations are inescapable.

The second factor, *information*, is common and essential in both hierarchies, too. On the farm we may be talking about a milk cow as an organism or about a garden, patch, field, farm, or farm community as ecosystems and as we do, we are at home in the hierarchy of structure. On the same farm we may be talking about the dairy herd, the hogs, or chickens, all local populations of species that are relatives of wild things and fall under the hierarchy of descent. The information stored in nearly every cell of an organism is responsible for its growth, development, daily living, and reproduction. The information packaged in each somatic cell is part of a cloned population of such information packages; the local dairy herd population is part of a global breed, which is related to all the domestic breeds and ultimately to wild cattle.

While all of this biological information is subject to Darwinian natural selection, cultural information is subject to Lamarckian selection (in which environmental changes cause structural changes in animals and plants). Acquired cultural information can be handed down; although it is more vulnerable than biological information as a whole, it is less vulnerable than most of us think. I heard recently that even though Taoism has no official status in China, it is alive in each village even today. Even so, less cultural information is present on the earth than biological information. To destroy biological information and expect cultural information to take its place on a one-to-one basis is risky. This is one reason the unsettling of America has been such a tragedy. Countless "bits" of cultural information have been lost in one generation.

The third factor common to both hierarchies is *energy*. Our natural world, assembled independently of human hands, is a world in which life, as the main or most interesting attraction, has had to rely almost exclusively on what we might call *contemporary sunlight*. The fact that sunlight is dispersed imposes certain limits on its use. The physical structures of the earth impose certain patterns of storage and transfer of energy. The various rocks and densities of air and water have their respective specific heats; a pound of air will hold about one-fifth as much heat as a pound of water. A pound of rock will hold a fifth to a fourth as much heat as a pound of water. Nature's information system accommodates all of these factors that influence the chemical interactions in the soil. It is staggering to contemplate, but this living world more than copes; it evolves and thrives. Except when an ecosystem uses fire, nature's biota operate at low temperatures.

By contrast, the industrial economy runs on energy that is not contemporary. It uses fossil fuels whose age must be measured in the tens of millions of years or nuclear fuels that are as old as the universe itself. In the use of all these forms of energy, high temperature is the rule. The technologies that respond best to such energy more than invite us to expand the scale of a farming operation. This is how the industrial economy has elbowed its way into agriculture, into processes that are fundamentally biological in nature. Scale expands in nearly every phase of the operation. We use fossil fuels to control our competitors—the weeds, insects, and pathogens. We compact the earth with heavy equipment powered by fossil fuels. Fertilizer is processed at high temperatures mostly with natural gas as the feedstock. If nature had not been so resilient, we might have backed out long before such a complete infrastructure was put in place to exploit highly concentrated, ancient energy. We might have started earlier to accommodate a more sensible agriculture, one that contains a biological, physical, and chemical structure tuned to contemporary solar energy. All biological information, whether it is in microbes, higher vertebrates, or flowering plants, is tuned to a sun-powered earth. When humans destroy this information, they must substitute cultural information, apply more energy, or do both. Usually much more energy is applied than cultural information, and ecological capital is wasted or polluted. Soil erodes and is treated with chemicals for fertility and pest control.

THE INTERACTION OF SCALE, INFORMATION, AND ENERGY

The importance to agriculture of understanding the three interacting dynamics common to both hierarchies may be best clarified when we contrast two ecosystems: a modern cornfield and a tall-grass prairie as an example of a natural ecosystem. If one could tease out the genetic instructions from a cell of a corn plant and type the instructions necessary to make the corn plant grow and prosper, the instructions would fill thousands of volumes, perhaps enough to fill a small house. If we could discover and record the biological information in all native prairie plants, it would fill all the libraries of the world and then some.

Biological information (stored in the DNA/RNA code of a myriad of life-forms) makes it possible for a native prairie ecosystem to remain intact without the human, provided all trophic levels that make up the prairie community are intact. Even where humans have turned native prairie into pasture for cattle by building fences and ponds, only a small amount of cultural information and intelligence is necessary because the biological information remains high. A ranch in the Flint Hills of eastern Kansas consists of nearly 6,400 acres of native prairie—nearly ten square miles. It has never been plowed and supports around 1,700 head of cattle during the grazing season. Aside from additional help needed for fencing, branding, castrating, loading, and unloading, that entire acreage is managed by a single cowboy who usually rides the range in a pickup truck. This is typical management practice for the region. Protecting that rolling country from erosion requires so little effort because the roots and leaves and small creatures of the soil of that diverse biota protect it and keep it stable. The roots are always there, as are the leaves, except when the prairie is burned. Biological information translates into chemical diversity and prevents epidemics of insects and pathogens, so that the cattle can have most of the grass. Ranchers never add fertilizer. They stock the land at the rate of about one cow per four acres. What prevents overstocking of that rangeland is the intelligence of the owner and the cultural information handed down. When signs of overstocking appear, the cattle are shipped out. The fossil energy input per acre in that prairie is practically nil. The ecosystem, powered by the sun's energy, provides its own fertility. Most of the weight of the cattle is water. The carbon and nitrogen ultimately come from the air. When the cattle are hauled off, the nutrients that came from the soil must be replaced by solar-powered root pumps extending into the subsoil or the rock a few inches to a few feet below the surface. They are the same mines that supported the American bison.

When we are talking about ecosystems being directed to meet the human purpose, the scale of operation becomes critical. When the biological information per acre is high, the cultural information being applied per acre can be very low. This means that the ratio of acres to humans can be very high as it is with our Flint Hills ranch. There are products of the industrial revolution on this prairie—the barbed wire, the steel fence posts, the small dams built with bulldozers, the pickup truck, even the breeding that has produced some of the highly efficient weight-gaining cattle. But overall, nature dominates this landscape. The fossil energy input per pound of protein, as well as the fossil energy input per acre, is very low. In this system, the ratio of biological information to cultural information per pound of produced protein may be the highest of any place on the land surface of the earth. As a food-producing ecosystem for humans, it is highly resilient, although much of the mammalian diversity is now gone.

On September 12, 1806, Zebulon Pike stood on a high hill just a few miles from this ranch and saw the prairie wolves, panthers, American bison, wapiti (elk), deer, and antelope. In less than one hundred years, white settlers eliminated their competitors in the food chain—the wolf, the panther, and the Native American—and the wild ungulate herbivores—the bison, wapiti, deer, and antelope—in favor of their own domestic ungulate, beef. In spite of these extirpations, the ranch ecosystem has remained essentially intact because humans and cattle are sufficiently close as ecological analogs to the larger wild grazers and their predators. Every lost and substituted creature was a mammal. Most of the prairie vegetation changed but little. But one important lesson remains: even in this instance, before ranchers could alter that wild ecosystem for their use, they had to simplify it.

Now let us move our attention to 6,400 acres of former prairie land in an Amish settlement in northern Indiana. With an average of eighty acres per Amish family, a diversified agriculture here would support eighty families. Even if the livestocking rate were four times as high, as I suspect it could have been, I doubt that management would require even four times the effort, time, and thought that our Flint Hills ranch requires. The point is, however, that if the soil and water are to be maintained where till agriculture is practiced, the jobs done by the prairie biota now must be done by humans and their domestic species. With the loss of biological diversity, the price for sustainability must be paid from elsewhere. The Amish substitute cultural information for biological information—muscle power from the draft animals and from humans. The Amish use small engines on some pieces of equipment, such as balers, which are pulled by draft animals. They also use tractors for belt power to run stationary equipment such as threshing machines. Consequently, the Amish may use more fossil fuel per acre of production than the Flint Hills ranch uses. Of course, they produce more food per acre, but they may not necessarily use fewer fossil-fuel calories to produce one thousand calories of food. Accurate comparisons are complicated because what the rancher harvests from the rangeland is mostly red meat, whereas the Amish farm produces grain and other plant products, draft power, milk, as well as meat. But for the Amish to farm the 6,400 acres, they must increase the ratio of people to acreage 150 to 200 fold. One reason for this increase is that with till agriculture, more attention must be paid to the vertical dimension, particularly below the surface.

The same northern Indiana acreage farmed with conventional equipment by industrial farmers would likely support fewer than half the number of families and probably one-third the number of people. (Amish families are larger than average.) The species diversity on the Amish farm is much higher than on the conventional farm—the Amish usually grow more crops and keep more kinds of animals. A larger measure of their fertility comes from crop rotation and manure. Many Amish farms do use commercial fertilizer and pesticides, but the extent of their use is usually much more restricted than on the average conventional farm.

In western Kansas, two related farm families may own and farm as many as 6,400 acres. Because this must be a dryland farming operation, they will probably devote most of their acreage to growing sorghum and wheat. There may be two generations associated with such an operation, or the families of two siblings may be farming together; there are many combinations. With seasonal outside help, this operation is viable for the present. Energy takes the place of information, and as long as the liquid fuel is available, farming this flat country can continue. Erosion is not a serious problem.

But what if such a 6,400-acre western Kansas farm were to use only the solar energy captured on that farm for field work, growth, and harvest? With a ratio of one draft animal to sixteen acres or more, the most energy-efficient way of doing the work would be to use the muscle power of draft animals and humans; with this ratio, if some of the grain or stover were turned into alcohol for tractor fuel, more energy would be required. Of course, if tractors were used, fewer people would be necessary. If all 6,400 acres were planted to short-grass prairie, it could still support livestock and be cared for in the same manner as our 6,400-acre Flint Hills ranch. Eighty acres wouldn't support even an Amish family in this area beginning around the 100th meridian (western Kansas), where evaporation is in excess of rainfall. Without irrigation, the portion of each year's crop that would have to go into the draft animals to grow the crop would be so great that the harvest available to sell could not be counted on to support the most frugal Amish family. Although a few Amish remain in Kansas, many more have moved. The Amish possess cultural information for farming in humid climates. Of those who have stayed, most farm with tractors.

But there is an important consideration raised by this comparison of the economies of northern Indiana and western Kansas. Even in Iowa, where there is an abundance of rainfall, a farm powered by draft animals will require more acres, although less energy, to support the animals than will be necessary to support tractors. Draft animals require a diet while tractors mostly require energy. To grow a balanced diet requires that we feature diversity, which forces crop rotations. In many cases, certain land now regarded as marginal or difficult can become permanent pasture. This may be land along a stream, which, on an industrial farm, is now "farmed right through." This kind of accounting forces us to realize that if we are to practice till agriculture without ruining the land, our standards have to be biological and not industrial. When the expectations of the land can't be met with conventional crops, the land should be returned to nature or to a vegetative structure that resembles that of nature.

SUMMARY

The ecosystem is a category immediately above organism in the hierarchy of objects from atoms through cells, tissues, and so on up. All the levels in the hierarchy obey the same laws of integration. Because these laws are common for every level, examples drawn from a lower level and applied to the ecosystem are of greater utility as an explanation than if they were mere analogies.

The hierarchy of ascent that we observe in the earth's biota is the consequence of speciation. Species can be viewed as information boundaries that, once established, were less information and energy entropic than their immediate ancestral species. The laws or rules governing this hierarchy have not been formulated.

Laws or rules govern the two hierarchies that are independent of one another, but the two are intricately intertwined and probably have been over the course of evolution. Common to these two intertwining hierarchies are three interacting dynamics: those of scale, information, and energy. Scale can expand or decrease, and as it does, the relationship between information and energy changes dramatically. For agriculturists, information is of two types: biological and cultural. Biological information is subject to Darwinian selection and is more reliable than cultural information, which is Lamarckian in nature, but both types of information are needed. There can be too much biological information in an ecosystem and on a farm. Too much diversity can spell instability, although this is seldom a problem on the farm. Most farms suffer from too much energy, too large a scale, and too little information. An abundance of energy, especially of the fossil-fuel variety, has a way of homogenizing the farm ecosystem. With too much energy, there is a strong incentive to reduce diversity in order to simplify the farm structure and, in so doing, to break down the complementarities in the biotic and cultural diversity on the farm.

Discovering the right balance of cultural and biological information and the balance between information and energy, given the scale of an operation, is necessary for sustainability, a synonym for homeostasis. It is out of an understanding of these dynamics that a taxonomy can begin to develop, a taxonomy that can comfortably include everything from a small garden to a large ranch.

We may finally come to understand all of this as a kind of harmony. In this respect, Wendell Berry has already summarized what I have said here:

> In a society addicted to facts and figures, anyone trying to speak for agricultural *harmony* is inviting trouble. The first trouble is in trying to say what harmony is. It cannot be reduced to facts and figures—though the lack of it can. It is not very visibly a function. Perhaps we can only say what it may be like. It may, for instance, be like sympathetic vibration: "The A string of a violin . . . is designed to vibrate most readily at about 440 vibrations per second: the note A. If that same note is played loudly not on the violin but near it, the violin A string may hum in sympathy." This may have a practical exemplification in the craft of the mud daubers which, as they trowel mud into their nest walls, hum to it, or at it, communicating a vibration that makes it easier to work, thus mastering their material by a kind of song. Perhaps the hum of the mud dauber only activates that anciently perceived likeness between all creatures and the earth of which they are made. For as common wisdom holds, like *speaks to* like.[13]

NOTES

1.   Arnold M. Schultz, "The Ecosystem as a Conceptual Tool in the Management of Natural Resources," in *Natural Resources: Quality and Quantity*, eds. S. V. Ciriacy Wantrup and James S. Parsons (Berkeley: University of California Press, 1967).

2. A. G. Tansley, "The Use and Abuse of Vegetational Concepts and Terms," *Ecology* 16 (1935): 284–307.
3. Cited in Schultz, "The Ecosystem as a Conceptual Tool," 141.
4. J. S. Rowe, "The Level of Integration Concept and Ecology," *Ecology* 42 (1961): 420–427.
5. J. K. Feibleman, "Theory of Integrative Levels," *British Journal of the Philosophy of Science* 5 (1954): 59–66.
6. In a comparison between old order Amish and corn belt farmers in northern Indiana, wildlife habitat interspersion was found to be three times greater on Amish farms. Smaller fields and three times as much edge are responsible for this much extra wildness. This is based on work by Mark Biggs, "Conservation Farmland Management—The Amish Family Farm Versus Modern Corn Belt Agriculture" (M.S. thesis, School of Forest Resources, Pennsylvania State University, May 1981).
7. P. W. Anderson, "More Is Different," *Science* 177 (1972): 393–396.
8. E. P. Odum, *Fundamentals of Ecology* (Philadelphia: W. B. Saunders, 1971), 5.
9. F. A. Bazzaz and J. A. D. Parrish, "Organization of Grassland Communities," in *Grasses and Grasslands: Systematics and Ecology*, eds. James Estes et al. (Norman: University of Oklahoma Press, 1982).
10. E. O. Wiley and Daniel R. Brooks, "Victims of History—A Nonequilibrium Approach to Evolution," *Systematic Zoology* 31, no. 1 (1982): 1–24.
11. R. M. May, *Stability and Complexity in Model Ecosystems* (Princeton, N.J.: Princeton University Press, 1973).
12. Wendell Berry, "An Agricultural Journal in Peru," in *The Gift of Good Land* (San Francisco: North Point Press, 1981).
13. Wendell Berry, "People, Land and Community," in *Standing by Words* (San Francisco: North Point Press, 1983), 76. Italics mine.

Toward a

Common Covenant

Probably no animal species alive has a greater variety of bioregions in its background than the human. This enormously varied evolutionary experience of our species accounts for our unspecialized nature, required for versatility, which contributes overall to our adaptability. These traits go far beyond making us adaptable in the ordinary sense. After all, we *create* environments in a way no other species has managed to do. In many cases, we have maintained these environments for centuries. The Netherlands is a good example of a region where some of us have claimed swampy mud flats from the sea. The people there have more than survived, for they have devised ways to make use of the abundant fertility of this low country to meet human demands and eventually to build a remarkable culture. More recently, our impressive cleverness has led us to locate a few men and women near the South Pole for several years, a few at the ocean's bottom for weeks, and two or three on or near the moon for days. It *is* an impressive feat when scientists can engineer a small earth to carry several persons on trips totalling more than half a million miles.

But none of these artificially created environments (and there are countless more) compares to our adventure into agriculture. Though agriculture involves an entirely different kind of sophistication, we ought still to compare it to other human ventures. The logician may avoid comparing apples and oranges or different *kinds* of sophistication, but in the real world, incommensurables do become commensurable, especially during periods of scarcity. I suspect that long before another half-century has gone by, we will be forced as a people to confront the fact that agriculture, overall, has been and now is many times more sophisticated a venture than space travel could ever become, completely towering over all other technological spectaculars. How spectacular something is has little, if anything, to do with its sophistication. Agriculture is more sophisticated partly perhaps because it comes out of living nature and is therefore complex, and partly because more human minds have worked on ways to generate an assured food supply than on any other task. The result has been that we have changed the face of the earth while providing meals for billions, all within four hundred human generations.

I am not dealing with this question of sophistication because agriculturists think of themselves as unsophisticated and need a pep talk, but rather to emphasize that the result of their handiwork dominates the global land mass so completely because we have used our *unspecialized* and versatile capabilities (which stand behind our sophistication) to meet our *specialized* demands from the earth, especially for our food. We depend almost exclusively on flowering plants, the last of earth's major plant types to evolve. Furthermore, within the entire spectrum of flowering plants is a very narrow band of plant types called grasses upon which we mostly depend. We plant most of the agricultural world in a few kinds of grasses, such as rice, wheat, and corn. Not only do we eat this specialized flora directly, as in our daily bread, we use it almost exclusively to support all our mammalian livestock and domestic fowl. Our grain crops are either annuals or treated as such and all are produced primarily in monoculture. When an unspecialized and versatile species makes such a specialized demand upon the environment, a split between humans and nature seems inevitable.

Perhaps, but only perhaps, a bioregional perspective will make a difference. Consider the political boundary between Montana and

Canada. Hasn't policy *and* culture made more of an impact here than the dictates of nature?

In short, I am not moved by some of the talk about bioregionalism. Instead, I believe earth has a problem primarily because we have a problem with our very nature, and no common covenant for governing or coping with that nature. The problem for the earth is that we have an unspecialized and versatile ability to meet some rather specialized demands.

For example, most of humanity depends on a few annual grass species grown in monoculture for most of its food. In most cases, the nature we have destroyed to grow food was more generalized than what we've replaced it with. Our versatile ability to homogenize environments comes from our experiences in numerous bioregions over time, experiences that our bodies and minds still remember. The African savannas applied some of the last and most important touches to this being of ours, but before the savanna experience there was the forest, a forest experience so old it has been left for our bodies to remember what our brains never comprehended. That's not completely true, of course, but we can observe some of this body memory every day of our life. We might observe it, for example, when we place a ten-penny nail to the surface of a two-by-four, when one of our freely articulating shoulders, derived from the forest, allows us to hold that nail high over our heads with one hand until the first tap with the hammer. With the other shoulder now in ratchet motion, the arm drives the nail home. The hammer is held by a clasping hand, first shaped to grasp limbs and, only later, stone tools and weapons. The forest that gave us that fine shoulder, and much of the hand, is the same forest that perfected our stereoscopic vision, the depth perception that grants us the justifiable confidence to hit that small nail and drive it all the way with a few solid strokes. It scarcely matters why it so satisfying to sink a nail that drives straight and true. But it *does* feel *so* right; I doubt that any other species could experience such satisfaction.

The satisfaction I have just described is part of a much larger matrix, a substrate with potential for both good and evil. The evil first appears when that part of our human nature, whether armed with the extra-human energy of an ox or a nuclear reactor or anything in between, threatens to reduce the earth's biological capital in order to *create* artificial environments specialized beyond necessity. Real trouble begins when the memories of our bodies and minds become our *desires*. It happens so unwittingly. To drive the nail straight and true and feel good about it is innocent enough. Your desires may be still adequately in check when your body lies awake at night tantalizing your mind, which in turn sees itself moving the body to take the pickup on Saturday morning to the lumberyard to buy two-by-fours, and a string level and bags of Portland cement and some sand and some nails and anything else necessary to start the ten-foot porch addition at the back. It all adds to our comfort and the household does benefit. So far, so good. But when more bodies and more minds, some trained as architects, some as engineers, see themselves building one more shopping center over one more wheat field, *then* we wish our planning commission had a "bioregional point of view."

We all want to do something to satiate *desire*, which in most cases goes far beyond need. Much that is responsible for these insatiable desires is derived from our experiences in countless bioregions of the past. I mentioned in another essay in this book that we are a "fallen species"—a species out of context. Without trying to excuse our waywardness, it does seem to me that *this* component of our original sin was born out of necessity. What is at work is obedience to the appetites of a body and mind shaped in a past in which there was no opportunity for a gluttonous consumption of resources or the fouling of the environment beyond redemption.

So we operate as educators in the faith that more knowledge will solve our problems. "Increase the volume of education," we say. But results are not guaranteed by simply increasing the *volume* of education without attending to content as well, as Aldo Leopold suggested over thirty years ago. Leopold thought we needed additional content to change our loyalties and affections. It is this we need to keep in mind in our search, for it forces us to confront the basic question of how much humans should try to improve on nature.

It's an old question, really, and thinkers through the ages have devoted lots of time to it. It goes way back. Remember that the emblem of this book is the altar of unhewn stone; and remember God's warning to refrain from polluting the stone of the altar with any tool. If the word *polluted* were not in the instructions, one might conclude that God wanted praise and he wanted it quick. Instead, we are forced to ask if God is not telling us that his creation is good enough as it is, or was, and that to modify his materials is to reduce their value. Or was the old jealous God simply worried that human cleverness and dexterity with a tool might shape a handiwork so dazzling that all who looked upon it would be more mindful of the artisan than of the Creator?

As an ecological preservationist, I have cited this scripture in support of the argument that to alter the original material is to neglect its larger purpose in the scheme of things. It is a tough problem. Would not even the most ardent rock hound quickly concede that the original chunk of marble that Michelangelo circled and studied for cracks was not nearly so grand as the finished *David*? The greatest wisdom has always been required to judge if nature acting alone can meet our needs, or if our cleverness and effort are required. Tampering with the environment is a serious matter. Animal and plant species improve their environments by reordering their surroundings. Birds build nests. Prairie dogs make tunnels. Plants change their shading patterns and in many cases exude chemicals from their roots that are toxic to their competitors. Ultimately, the question must have to do with the *time necessary for recovery* after an environment has been overly modified.

God had a problem, for, as I have suggested earlier, the human impulse to create has had too much survival leverage in our gathering-hunting past. One can't expect us to override easily this psycho-biological drive, even when the orders are from above and are explicit.

FRAGMENTED VISIONS

With this as background, consider the recent mania among scientists to alter or rearrange the DNA of numerous species to serve human *needs* and *desires*. Frankly, I am not concerned about some "cut loose" genes from one species causing a recipient plant to run wild when "stitched in" by some clever biochemist. I won't be surprised if we eventually find human genes in most of our domestic crops, genes that long ago were surgically removed by viruses and carried to the DNA of a crop plant and finally stitched in with the aid of the viral surgeon. Such plants would have a selective advantage over their noninfected neighbors. Such *co*evolution may have been going on for ten to twenty thousand years.

My bet is that nothing fundamentally new, biologically, is actually underway. The problem with DNA surgery, gene splicing, genetic stitching, or whatever, is not that it might possibly upset the balance of nature, but that it will almost certainly further upset much of whatever is left of the balance of the human. It is once more an emphasis on cleverness that is motivated by the desire for a "breakthrough." What follows a "breakthrough" may best be characterized as a spasm. The likes of the *Wall Street Journal*, *New York Times*, *Washington Post*, and *San Francisco Chronicle* will devote several column inches to it. The scientific "sprinters" who brought it about will suddenly be invited to more parties than they can possibly attend.

What if some clever biochemists do manage a "breakthrough" so that sunflowers and grasses fix nitrogen as actively as certain legumes, so that we can momentarily keep the famines from trimming the human population in a part of the world where overcrowding is already terrible? Even a casual study of the history of science and technology forces us to admit that, in most cases, the wrong solution of one problem contributes to the creation of more problems. I am not talking about the old saw that more people surviving means that more people breed, upping the ante one more time. I am talking about the arrogance that attends our seeing the world in parts rather than as a whole. A recent example of what can happen is the fever following the discovery of interferon, that substance thought to be one of our body's best natural defenses against viral infections. Millions were spent by companies to manufacture the substance through DNA transfer and the like before it was discovered that interferon actually increases the damage done in mice by yet other viruses. They may get the entire problem solved eventually, but how far and how extensively is it appropriate for us to noodle around in the nucleus? As much as we want and can? Most people seem to think so, but we can guess what God would say.

Only the fool or the complete nihilist would have anything to do with a proposal to stop all art and science. As I see it, our problem is that we lack a common covenant and a common vision to guide us. So our scientists, with few exceptions, are free to do what is possible. The people Moses ordered around lived under a common covenant and had a clear idea of where they were going and why. Moses had plenty of trouble on his hands, even so. Centuries later, that covenant was still very powerful when David, Israel's second king, got completely out of line by bringing Bathsheba into his palace, impregnating her, and having her husband killed in a battle. What David had done was quickly recognized as unacceptable. The prophet Nathan appeared and told David a parable. When David became outraged about the injustice done in the story, Nathan quickly put the finger on David by saying, "Thou art the man." Because they lived under a common covenant, the prophet made it stick. King David repented and suffered a terrible consequence—the love child died. The prophetic tradition has always carried the best of the people and has always stood against the royal tradition.

AWARENESS AND PENANCE

I will come back to the idea of a common covenant later, for it ties in with another religious idea, the idea of atonement. The Hebrew idea of atonement isn't taken very seriously by many people anymore, even though the featured animal, the scapegoat, is frequently mentioned. The scapegoat was the animal over whose forehead the sins of the people were recited before it was set free to wander in the wilderness. Actually, there are two goats associated with atonement: one, which was sacrificed, symbolized the forgiveness of the people's sins; a second, which was left alive to wander from the wilderness back to the human habitat every now and then, into the view of the people, providing them with the occasional reminder that their sins, though forgiven, live on, or at least that the consequences of their sins live on. Even if they never saw the goat again, they lived with the knowledge that it wandered the wilderness bearing their sins.

But we can't talk about atonement unless we talk about what we are atoning for. Usually we think of atoning for our sins, for the evil we do on purpose. However, we have to think too about atoning for our *errors*—a cropping system which leads to terrible erosion, for example. The distinction between evil and error is not always clear, and this isn't the place to explore the distinction. What is important is that error is likely to require as much effort for rectification as evil. Redemption can follow either, for redemption is an ecological reality that must very early have become a religious notion, for whether ignorance or greed leads to overgrazing of a hillside, the result is the same. The hillside can be redeemed, probably not completely, probably not very soon, but eventually and to some degree.

Whether for our sins or our mistakes, we need a kind of ecological atonement, but no modern parallels for either of the two goats come to mind. One reason is that this Hebrew ritual of atonement involves sins committed on purpose, with a consciousness that everything the sinners had confessed was wrong. Many, if not most, of our sins against the land and its community are unconscious. We're not even aware that some of our acts result from arrogance.

We have an additional problem: when we try to correct our mistakes, we almost inevitably make more. And as we compound our errors, as Conn Nugent says, we should acknowledge those events or technologies which have been responsible. Some are outright failures. It would be convenient if we had an animal's forehead over which to recite our errors, along with our sins—ecological goats to wander in and out of our lives.

We might feel silly engaging in rituals that involve goats, living or dead, but there is nothing frivolous about the value of atonement or the need for a common covenant. A common covenant can unite a people around appropriate "loyalties and affections" and affirm their efforts to develop a healthy relationship among themselves and between them and nature. A sustainable agriculture will require both, for without both, essays on bioregionalism will be little more than lessons in geography and natural history.

What if we employed our rivers and creeks in some ritual of atonement? Their sediment load is largely the result of agricultural practices based upon arrogance, tied in turn to an economic system based upon arrogance. Much of that sediment load results from ignorance and error but too much more is the consequence of deliberate evil. I can't visualize what the ritual would look like, but perhaps we need an annual formal observance in the spring—when the rivers are particularly muddy—a kind of ecological rite of atonement, in which we would "gather at the river." Maybe we should ally ourselves by virtue of a common watershed, rather than a common ecosystem or bioregion. This is an important distinction, for a watershed can and often does cut through more than one bioregion. There would be nothing abstract about a common covenant among people of a common watershed. A cattle crossing a mile upstream from a town's water supply affects the amount of chlorine people will be forced to drink. A stream has as much potential to unify as the convenant shared by King David and the prophet Nathan.

A stream is a powerful image, too. It unites every square foot of its watershed. People of a common watershed, committed to a sustainable agriculture, must above all else recognize their future and that of their descendents in every square foot of that watershed. Anyone interested in maintaining the quantity of nutrients and the quality or health of a soil can judge the relative success or failure of those ideals in the river. Such a judgment will not suit the simple-minded, for rivers will always carry some sediment. They always have. Banks will slip, a river will cut. Rivers will continue to do what they did before agriculture, before humans. But in the long run, the sediment load of a river should not exceed the net accumulation of soil on the surface of the watershed. Root pumps serve both to hold the soil in place and to retain its nutrients. Root pumping is a form of mining, but roots also *retain* most of what they've mined. Our major sin as white occupiers of this continent, a sin that ranks with our enslavement of the African and our systematic genocide of the Native Americans, is the violence we've done to that *land community* of native plants and animals, the only assured guardian against the seaward flush of hard-earned soil to a watery grave.

WATER

When we think about water very extensively, we are forced to think globally. Someone worked it out once that every molecule of water at one time or another has been positioned on or above every square foot of earth. Moreover, water anywhere, but perhaps most particularly clear water running in a river, can remind us of what a special place earth is. Consider the fact that liquid water is comparatively rare in outer space where temperatures range from near absolute zero to the unimaginable heat in the cores of younger stars. We take liquid water for granted, a conceit by galactic standards, for under our atmospheric pressure it exists only within the extremely narrow temperature range of 180°—between 32°F. and 212°F. Everywhere else in the universe, where water exists at all, it must be either solid ice or vapor. Its liquid properties even for our own solar system, let alone this part of the sidereal universe, are unique.

But just because water is a liquid here on earth, we are not home free. We are land animals, and water's behavior on land affects us. On land, water has the potential to be either a life-nurturing source or a primary eroding force. Water, more dramatically than any of our other necessities, increases or diminishes human options. What if we, as a people, should actually visualize the journey of a water molecule in a raindrop from the time the atmosphere releases it to journey through the biota, until it returns again to one of its more permanent homes, the ocean or the atmosphere? We would soon become aware that the traps and gates that stall and shunt the itinerant water molecules are the most elaborate in nature's ecosystems.

This is what the human inhabitants of a watershed must know and understand if they are to "gather at the river." With practice, we can learn to read the signs that indicate a *tolerable* rate of sediment loss. When we see a load beyond this tolerable rate, we will realize that the consequences of the sins of the people (maybe all our sins if we take the argument far enough) have literally been cast out upon the waters for all to see. In a way, the river is our living goat. In another way, the river is better than a scapegoat, for it will respond when we do better, revealing our improvement almost daily by its increasing clarity.

Our sins against the land on behalf of agriculture alone are partly the consequences of power far beyond the length and breadth of any one watershed. Export policy is tied to balance-of-payment considerations to purchase foreign oil for an energy-wasting society. The institutional structure of American agriculture tempts farmers to overcapitalize, to bulldoze out Roosevelt-era shelterbelts, to plow fencerow to fencerow, to break out marginal land which had been kept in grass. Even though many of the sins against the land now stretch beyond the control of the individual or the boundaries of any watershed, the watershed still has the potential to unify. The river will always speak a language that humans can learn with a little effort—a language in every way local, in every way universal, but in no way foreign.

Yahweh's people had just shaken the paradigm of slavery and were thinking through another when he handed down the order to make the altar of unhewn stone. We know that in other lands, in ancient times before Yahweh had been named, sacred groves were set aside by others who had concerned themselves with the destruction of an Asian countryside, and even though we recognize the paradox that many of those groves were later cut by the faithful to build temples, we cannot imagine how much worse it would have been without those trees as reminders. If a "voice" were to give ecological instructions to prairie people today on how partially to atone for our sins and errors, I imagine it would say:

> Return as much of your land *as you can afford* to diverse native prairie. Do not add improved varieties that are the products of the tools of modern technology, lest you pollute the landscape. Do not try to improve on this patch of native prairie, for it will serve as your standard by which to judge your agricultural practices. There is no higher standard of your performance than the land and its natural community.